Stable Domination and Independence in Algebraically Closed Valued Fields

This book addresses a gap in the model-theoretic understanding of valued fields that has, until now, limited the interactions of model theory with geometry. It contains significant developments in both pure and applied model theory.

Part I of the book is a study of stably dominated types. These form a subset of the type space of a theory that behaves in many ways like the space of types in a stable theory. This part begins with an introduction to the key ideas of stability theory for stably dominated types. Part II continues with an outline of some classical results in the model theory of valued fields and explores the application of stable domination to algebraically closed valued fields. The research presented here is made accessible to the general model theorist by the inclusion of the introductory sections of each part.

Deirdre Haskell is a Professor in the Department of Mathematics and Statistics at McMaster University.

Ehud Hrushovski is a Professor in the Department of Mathematics at the Hebrew University of Jerusalem.

Dugald Macpherson is a Professor in the School of Mathematics at the University of Leeds.

LECTURE NOTES IN LOGIC

A Publication of
The Association for Symbolic Logic

This series serves researchers, teachers, and students in the field of symbolic logic, broadly interpreted. The aim of the series is to bring publications to the logic community with the least possible delay and to provide rapid dissemination of the latest research. Scientific quality is the overriding criterion by which submissions are evaluated.

See end of book for a list of the books in the series. More information can be found at http://www.aslonline.org/books-lnl.html.

LECTURE NOTES IN LOGIC 30

Stable Domination and Independence in Algebraically Closed Valued Fields

DEIRDRE HASKELL
McMaster University

EHUD HRUSHOVSKI
Hebrew University

DUGALD MACPHERSON
University of Leeds

ASSOCIATION FOR SYMBOLIC LOGIC

CAMBRIDGE
UNIVERSITY PRESS

CAMBRIDGE UNIVERSITY PRESS
Cambridge, New York, Melbourne, Madrid, Cape Town,
Singapore, São Paulo, Delhi, Tokyo, Mexico City

Cambridge University Press
32 Avenue of the Americas, New York, NY 10013-2473, USA

www.cambridge.org
Information on this title: www.cambridge.org/9780521335157

Association for Symbolic Logic
David Marker, Publisher
Department of Mathematics, Statistics, and Computer Science (M/C249)
University of Illinois at Chicago
851 S. Morgan St.
Chicago, IL 60607, USA
www.aslonline.org

© Association for Symbolic Logic 2008

First published 2008
First paperback edition 2011

A catalogue record for this publication is available from the British Library

Library of Congress Cataloguing in Publication data

Haskell, Deirdre, 1963–
Stable domination and independence in algebraically closed valued fields /
Deirdre Haskell, Ehu Hrushovski, Hugh Dugald Macpherson.
p.   cm. – (Lecture notes in logic : 30)
Includes bibliographical references and index.
ISBN 978-0-521-88981-0 (hardback)
1. Model theory.   2. Valued fields.   3. Domination (Graph theory)
I. Hrushovski, Ehud, 1959– II. Macpherson, Dugald. III. Title. IV. Series.
QA9.7.H377 2007
511.3′4 – dc22   2007036704

ISBN 978-0-521-88981-0 Hardback
ISBN 978-0-521-33515-7 Paperback

CONTENTS

PREFACE   ix

CHAPTER 1.   INTRODUCTION   1

PART 1.   STABLE DOMINATION . . . .

CHAPTER 2.   SOME BACKGROUND ON STABILITY THEORY   13
   2.1.   Saturation, the universal domain, imaginaries   15
   2.2.   Invariant types   17
   2.3.   Conditions equivalent to stability   18
   2.4.   Independence and forking   20
   2.5.   Totally transcendental theories and Morley rank   23
   2.6.   Prime models   23
   2.7.   Indiscernibles, Morley sequences   24
   2.8.   Stably embedded sets   25

CHAPTER 3.   DEFINITION AND BASIC PROPERTIES OF St_C   27

CHAPTER 4.   INVARIANT TYPES AND CHANGE OF BASE   41

CHAPTER 5.   A COMBINATORIAL LEMMA   53

CHAPTER 6.   STRONG CODES FOR GERMS   59

PART 2.   INDEPENDENCE IN ACVF . . . .

CHAPTER 7.   SOME BACKGROUND ON ALGEBRAICALLY CLOSED VALUED
            FIELDS   69
   7.1.   Background on valued fields   69
   7.2.   Some model theory of valued fields   71
   7.3.   Basics of ACVF   72
   7.4.   Imaginaries, and the ACVF sorts   73
   7.5.   The sorts internal to the residue field   77
   7.6.   Unary sets, 1-torsors, and generic 1-types   78

7.7.   One-types orthogonal to Γ 83
7.8.   Generic bases of lattices 85

CHAPTER 8.   SEQUENTIAL INDEPENDENCE 87

CHAPTER 9.   GROWTH OF THE STABLE PART 99

CHAPTER 10.   TYPES ORTHOGONAL TO Γ 103

CHAPTER 11.   OPACITY AND PRIME RESOLUTIONS 115

CHAPTER 12.   MAXIMALLY COMPLETE FIELDS AND DOMINATION 123

CHAPTER 13.   INVARIANT TYPES 137
13.1.   Examples of sequential independence 137
13.2.   Invariant types, dividing and sequential independence 143

CHAPTER 14.   A MAXIMUM MODULUS PRINCIPLE 151

CHAPTER 15.   CANONICAL BASES AND INDEPENDENCE GIVEN BY MODULES   161

CHAPTER 16.   OTHER HENSELIAN FIELDS 171

REFERENCES 177

INDEX 181

PREFACE

Valuations are among the fundamental structures of number theory and of algebraic geometry. This was recognized early by model theorists, with gratifying results: Robinson's description [45] of algebraically closed valued fields as the model completion of the theory of valued fields; the Ax-Kochen, Ershov study of Henselian fields of large residue characteristic with the application to Artin's conjecture [1, 2, 3, 10]; work of Denef and others on integration; and work of Macintyre, Delon, Prestel, Roquette, Kuhlmann, and others on p-adic fields and positive characteristic. The model theory of valued fields is thus one of the most established and deepest areas of the subject.

However, precisely because of the complexity of valued fields, much of the work centers on quantifier elimination and basic properties of formulas. Few tools are available for a more structural model-theoretic analysis. This contrasts with the situation for the classical model complete theories, of algebraically closed and real closed fields, where stability theory and o-minimality make possible a study of the category of definable sets. Consider for instance the statement that fields interpretable over $\mathbb{C}$ are finite or algebraically closed. Quantifier elimination by itself is of little use in proving this statement. One uses instead the notion of ω-stability; it is preserved under interpretation, implies a chain condition on definable subgroups, and, by a theorem of Macintyre, ω-stable fields are algebraically closed. With more analysis, using notions such as generic types, one can show that indeed every interpretable field is finite or definably isomorphic to $\mathbb{C}$ itself. This method can be extended to differential and difference fields. Using a combination of such methods and of ideas of manifolds and Lie groups in a definable setting, Pillay was able to prove similar results for fields definable over $\mathbb{R}$ or $\mathbb{Q}_p$. But just a step beyond, a description of interpretable fields seems out of reach of the classical methods. When p-adic or valuative geometry enters in an essential way, an intrinsic analog of the notion of generic type becomes necessary.

For another example, take the notion of connectedness. In stability, no topology is given in advance, but one manages to define stationarity of types or connectedness of definable groups, by looking at the type space. In o-minimality, natural topologies on definable sets exist, and connectedness, defined in terms of *definable paths*, is a central notion. In valued fields, the valuation topology is analogous to the o-minimal order topology, and one has the linearly ordered value group that may serve as the domain of a path; but every continuous definable map from the value group to the field is constant, and a model-theoretic definition of connectedness is missing. The lack of such structural model-theoretic understanding of valued fields is a central obstacle to a wider interaction of model theory with geometry in general.

It is this gap that the present monograph is intended to address. We suggest an approach with two components. We identify a certain subset of the type space, the set of *stably dominated types*, that behaves in many ways like the types in a stable theory. Since these types are not literally stable, it is necessary to first describe abstractly an extension of stability theory that includes them. Secondly, we note the existence of o-minimal families of stably dominated types and show, at least over sufficiently rich bases, that any type can be viewed as a limit of such a family. This requires imaginaries in a concrete form, serving as canonical bases of stably dominated types, and a theory of definable maps from Γ into such sets of imaginaries. Thus, whereas type spaces work best in stable theories, and definable sets and maps in o-minimal theories, we suggest here an approach mixing the two. As both the method and the intended applications depend heavily on imaginary elements, we develop some techniques for dealing with them, including prime models that often allow a canonical passage from imaginary to real bases.

We work throughout with the model completion, algebraically closed valued fields. This is analogous to Weil's program of understanding geometry first at the level of algebraic closure. One can hope that the geometry of other valued fields could also, with additional work, be elucidated by this approach. As an example of this viewpoint, consider the known elimination of quantifiers for Henselian fields of residue characteristic zero, relative to the value group and residue field. This was originally derived as an independent theorem. But it is also an immediate consequence of a fact about algebraically closed valued fields of characteristic zero, namely that over any subfield F, any F-definable set is definably isomorphic to pullbacks of definable subsets of the residue field and value group. See [19] for more details on this short argument. It is also noted there that definable sets in F^n can be fibered over the residue field and value group in fibers that are ACVF-definable, and a similar statement for definable bijections between them can be made. Another example is the elimination of imaginaries for $\mathbb{Q}_p$, proved in [20] using the ACVF methods of invariant types. We illustrate the phenomenon in Chapter 16 using the theory of $\mathbb{C}(t)$.

We restrict ourselves in this monograph to laying the foundations of this approach. Definable maps from Γ into imaginaries were described in [12]; we concentrate here on the theory of stable domination. Only future work based on these foundations can show to what extent they are successful. We note here that Pillay has defined an analogue of stable domination, *compact domination*, that appears to be useful for thinking about o-minimal groups (cf. [21]). Some progress has been made with the analysis of definable groups using the present methods; a notion of *metastability* has been abstracted (Definition 4.11), and results obtained for Abelian groups in a general metastable setting, and linear groups interpretable in ACVF; see [15]. And in very recent work of one of the authors with Loeser, connections with the Berkovich theory of rigid analytic spaces are beginning to emerge.

Acknowledgments

This work has benefitted from discussions over ten years with many mathematicians, and it is hard to single out individuals. The research of Haskell was supported by NSF and NSERC grants, and that of Hrushovski was partially conducted while he was a Clay Mathematics Institute Prize Fellow and was also partially supported by the Israel Science Foundation. The work of Macpherson was supported by EPSRC and LMS grants. All three authors were supported by NSF funding for parts of a semester at the Mathematical Sciences Research Institute, Berkeley.

CHAPTER 1

INTRODUCTION

As developed in [49], stability theory is based on the notion of an *invariant* type, more specifically a *definable type*, and the closely related theory of *independence of substructures*. We will review the definitions in Chapter 2 below; suffice it to recall here that an (absolutely) invariant type gives a recipe yielding, for any substructure A of any model of T, a type $p|A$, in a way that respects elementary maps between substructures; in general one relativizes to a set C of parameters, and considers only A containing C. Stability arose in response to questions in pure model theory, but has also provided effective tools for the analysis of algebraic and geometric structures. The theories of algebraically and differentially closed fields are stable, and the stability-theoretic analysis of types in these theories provides considerable information about algebraic and differential-algebraic varieties. The model companion of the theory of fields with an automorphism is not quite stable, but satisfies the related hypothesis of simplicity; in an adapted form, the theory of independence remains valid and has served well in applications to difference fields and definable sets over them. On the other hand, such tools have played a rather limited role, so far, in o-minimality and its applications to real geometry.

Where do valued fields lie? Classically, local fields are viewed as closely analogous to the real numbers. We take a "geometric" point of view however, in the sense of Weil, and adopt the model completion as the setting for our study. This is Robinson's theory ACVF of algebraically closed valued fields. We will view valued fields as substructures of models of ACVF. Moreover, we admit other substructures involving imaginary elements, notably codes for lattices; these have been classified in [12]. This will be essential not only for increasing the strength of the statements, but even for formulating our basic definitions.

A glance at ACVF reveals immediately a stable part, the residue field k; and an o-minimal part, the value group Γ. Both are stably embedded, and have the induced structure of an algebraically closed field, and an ordered divisible abelian group, respectively. But they amount between them to a small part of the theory. For instance, over the uncountable field $\mathbb{Q}_p$, the residue field has only finitely many definable points, and both k and Γ are countable in the model $\mathbb{Q}_p^a$. As observed by Thomas Scanlon [46], ACVF is not stable over Γ, in the sense of [48].

We seek to show nevertheless that stability-theoretic ideas can play a significant role in the description of valued fields. To this end we undertake two logically independent but mutually motivating endeavors. In Part I we introduce an extension of stability theory. We consider theories that have a stable part, define the notion of a *stably dominated* type, and study its properties. The idea is that a type can be controlled by a very small part, lying in the stable part; by analogy, (but it is more than an analogy), a power series is controlled, with respect to the question of invertibility for instance, by its constant coefficient. Given a large model $\mathcal{U}$ and a set of parameters C from $\mathcal{U}$, we define St_C to be a many-sorted structure whose sorts are the C-definable stably embedded stable subsets of the universe. The basic relations of St_C are those given by C-definable relations of $\mathcal{U}$. Then $\mathrm{St}_C(A)$ (the stable part of A) is the definable closure of A in St_C. We write $A \underset{C}{\overset{d}{\downarrow}} B$ if $\mathrm{St}_C(A) \downarrow \mathrm{St}_C(B)$ in the stable structure St_C and $\mathrm{tp}(B/C\,\mathrm{St}_C(A)) \vdash \mathrm{tp}(B/CA)$, and say that $\mathrm{tp}(A/C)$ is *stably dominated* if, for all B, whenever $\mathrm{St}_C(A) \downarrow \mathrm{St}_C(B)$, we have $A \underset{C}{\overset{d}{\downarrow}} B$. In this case $\mathrm{tp}(A/\operatorname{acl}(C))$ lifts uniquely to an $\operatorname{Aut}(\mathcal{U}/\operatorname{acl}(C))$-invariant type p. Base-change results (under an extra assumption of existence of invariant extensions of types) show that if p is also $\operatorname{Aut}(\mathcal{U}/\operatorname{acl}(C'))$-invariant then $p|C'$ is stably dominated; hence, under this assumption, stable domination is in fact a property of this invariant type, and not of the particular base set. We formulate a general notion of domination-equivalence of invariant types (2.2). In these terms, an invariant type is stably dominated iff it is domination-equivalent to a type of elements in a stable part St_C.

Essentially the whole forking calculus becomes available for stably dominated types. Properties such as definability, symmetry, transitivity, characterization in terms of dividing, lift easily from St_C to $\downarrow^d$. Others, notably the descent part of base change, require more work and in fact an additional assumption: that for any algebraically closed substructure $C \subseteq M \models T$, any type p over C extends to an $\operatorname{Aut}(M/C)$-invariant type p' over M.

We isolate a further property of definable types in stable theories. Two functions are said to have the same *germ* relative to an invariant type p if they agree generically on p. In the o-minimal context, an example of this is the germ at ∞ of a function on $\mathbb{R}$. Moving from the function to the germ one is able to abstract away from the artifacts of a particular definition. In stability, this is an essential substitute for a topology. For instance, if f is a function into a sort D, one shows that the germ is internal to D; this need not be the case for a code for the function itself. In many stable applications, the strength of this procedure depends on the ability to reconstruct a representative of the germ from the germ alone. We say that a germ is *strong* if this is the case.

It is easy to see the importance of strong germs for the coding of imaginaries. One wants to code a function; as a first approximation, code the germ of the function; if the code is strong, one has succeeded in coding at least a (generic) piece of the function in question. If it is not, one seems to have nothing at all.

We show that germs of stably dominated invariant types are always strong. The proof depends on a combinatorial lemma saying that finite set functions on pairs, with a certain triviality property on triangles, arise from a function on singletons; in this sense it evokes a kind of primitive 2-cohomology, rather as the fundamental combinatorial lemma behind simplicity has a feel of 2-homology. Curiously, both can be proved using the fundamental lemma of stability.

In [15] is is shown that stable domination works well with definable groups. A group G is called generically metastable if it has a translation invariant stably dominated definable type. In this case there exists a unique translation invariant definable type; and the stable domination can be witnessed by a definable homomorphism $h : G \to H$ onto a connected stable definable group. Conversely, given such a homomorphism h, G is generically metastable iff the fiber of h above a generic element of H is a complete type. Equivalently, for any definable subset R of G, the set Y of elements $y \in H$ such that $h^{-1}(y)$ is neither contained in, nor disjoint from R is a small set; no finite union of translates of Y covers H. We show this in Theorem 6.13, again using strong germs.

The general theory is at present developed locally, at the level of a single type. It is necessary to say when we expect it to be meaningful globally. The condition cannot be that every type be stably dominated; this would imply stability. Instead we would like to say that uniformly definable families of stably dominated types capture, in some sense, all types. Consider theories with a distinguished predicate Γ, that we assume to be linearly ordered so as to sharply distinguish it from the stable part. We define a theory to be *metastable over* Γ (Definition 4.11) if every type over an algebraically closed set extends to an invariant type, and, over sufficiently rich base sets, every type falls into a Γ-parameterized family of stably dominated types. We show that this notion is preserved under passage to imaginary sorts.

The proviso of "sufficiently rich base set" is familiar from stability, where the primary domination results are valid only over sufficiently saturated models; a great deal of more technical work is then needed to obtain some of them over arbitrary base. The saturation requirement (over "small" base sets) is effective since types over a model are always based on a small set. In the metastable context, more global conditions incompatible with stability are preferred. This will be discussed for ACVF below.

For some purposes, extensions of the base are harmless and the theory can be used directly. This is so for results asserting the existence of a canonical definable set or relation of some kind, since a posteriori the object in question is defined without extra parameters. This occurred in the classification of maps from Γ in [12]. Another instance is in [15], where under certain finiteness of rank assumptions, it is shown that a metastable Abelian group is an extension of a group interpretable over Γ by a definable direct limit of generically metastable groups.

In Part II we study ACVF. This is a C-minimal theory, in the sense of [35], [13]: there exists a uniformly definable family of equivalence relations, linearly ordered by refinement; their classes are referred to as (ultrametric) *balls*; and any definable set (in 1-space) is a Boolean combination of balls. In strongly minimal and o-minimal contexts, one often argues by induction on dimension, fibering an n-dimensional set over an $n-1$-dimensional set with 1-dimensional fibers, thus reducing many questions to the one-dimensional case over parameters. This can also be done in the C-minimal context. Let us call this procedure "dévissage".

A difficulty arises: many such arguments require canonical parameters, not available in the field sort alone. And certainly all our notions, from algebraic closure to stable embeddedness, must be understood with imaginaries. The imaginary sorts of ACVF were given concrete form in [12]: the spaces S_n of n-dimensional lattices, and certain spaces T_n, fibered over S_n with fibers isomorphic to finite dimensional vector spaces over the residue field. But though concrete, these are not in any sense one-dimensional; attempting to reduce complexity by induction on the number of coordinates only leads to subsets of S_n, which is hardly simpler than $(S_n)^m$.

Luckily, S_n itself admits a sequence of fibrations $S_n = X_N \to X_{N-1} \to \cdots \to X_0$, with X_0 a point and such that the fibers of $X_{i+1} \to X_i$ are o- or C-minimal. This uses the transitive action of the solvable group of upper triangular matrices on S_n; see the paragraph following Proposition 7.14. There is a similar statement for T_n (where strongly minimal fibers also occur.) It follows that any definable set of imaginaries admits a sequence of fibrations with successive fibers that are strongly, o- or C-minimal ("unary sets"), or finite. Dévissage arguments are thus possible.

One result obtained this way is the existence of invariant extensions. A type over a base set C can only have an invariant extension if it is *stationary*, i.e., implies a complete type over acl(C). We show that in ACVF, every stationary type over C has an Aut($\mathcal{U}/C$)-invariant extension. For C-minimal sets (including strongly minimal and o-minimal ones), there is a standard choice of invariant extension: the extension avoiding balls of radius smaller than necessary.

But this does not suffice to set up an induction, since for finite sets there is no invariant extension at all. Thus a minimal step of induction consists of *finite covers of C-minimal sets*, i.e., with sets Y admitting a finite-to-one map

$\pi\colon Y \to X$, with X unary. This is quite typical of ACVF, and resembles algebraic geometry, where dévissage can reduce as far as *curves* but not to a single variable. In the o-minimal case, by contrast, one can do induction on ambient dimension, or the number of coordinates of a tuple; this explains much of the more "elementary" feel of basic o-minimality vs. strong minimality.

The additional ingredient needed to obtain invariant extensions of types is the *stationarity lemma* from [12], implying that if π admits a section over a larger base, then it admits a section over $\mathrm{acl}(C)$. See Lemma 8.10. For the theory ACF over a perfect field, stationarity corresponds to the notion of a regular extension, and the stationarity lemma to the existence of a geometric notion of irreducibility of varieties. It is instructive to recall the proof for ACVF. Given a finite cover $\pi\colon Y \to X$ as above, a section s of Y will have a strong germ with respect to the canonical invariant extension of any type of X. Generic types of closed balls are stably dominated; for these, by the results of Part I, all functions have strong germs. Other types are viewed as limits of definable maps from Γ into the space of generics of closed balls. For instance if $\tilde{b}$ is an open ball, consider the family of closed sub-balls b of $\tilde{b}$; these can be indexed by their radius $\gamma \in \Gamma$ the moment one fixes a point in $\tilde{b}$; by the stably dominated case, one has a section of π over each b. The classification of definable maps from Γ (actually from finite covers of Γ) is then used to glue them into a single section, over the original base. This could be done abstractly for C-minimal theories whose associated (local) linear orderings satisfy $\mathrm{dcl}(\Gamma) = \mathrm{acl}(\Gamma)$. The proof of elimination of imaginaries itself has a similar structure; see a sketch at the end of Chapter 15.

Another application of the unary decomposition is the existence of canonical resolutions, or prime models. In the field sorts, ACVF has prime models trivially; the prime model over a nontrivially valued field F is just the algebraic closure F^{alg}. In the geometric sorts the situation becomes more interesting. The algebraic closure does not suffice, but we show that finitely generated structures (or structures finitely generated over models) do admit canonical prime models. A key point is that the prime model over a finitely generated structure A add to A no elements of the residue field or value group. This is important in the theory of motivic integration; see the discussion of resolution in [19]. A further application of canonical resolution is a quantifier-elimination for $\mathbb{C}((t))$ in the $\mathcal{G}$-sorts, relative to the value group Γ. In essence resolution is used to produce functions on imaginary sorts; in fact for any $\mathcal{G}$-sort represented as X/E and any function h on X into the value group or residue field, there exists a function H on X/E such that $H(u) = h(x)$ for some $x \in X/E$.

The construction of prime models combines the decomposition into unary sets with the idea of opacity. An equivalence relation E on X is called *opaque* if any definable subset of X is a union of classes of E, up to a set contained in finitely many classes. This is another manifestation of a recurring theme.

Given $f: X \rightarrow Y$ and an ideal I on Y, we say X is dominated by Y via (f, I) if for any subset R of X, for I-almost every $y \in Y$, the fiber $f^{-1}(y)$ is contained in R or is disjoint from it. For stable domination, Y is stable and I is the forking ideal; for stationarity, f has finite fibers, and I is the dual ideal to an invariant type; for opacity, I is the ideal of finite sets. The equivalence relations associated with the analyses of S_n and T_n above are opaque. For an opaque equivalence relation, all elements in a non-algebraic class have the same type (depending only on the class); this gives a way to choose elements in such a non-algebraic class canonically up to isomorphism. Algebraic classes are dealt with in another way.

We now discuss the appropriate notion of a "sufficiently rich" structure. In the stable part, saturation is the right requirement; this will not actually be felt in the present work, since the stable part is $\aleph_1$-categorical and does not really need saturation. For the o-minimal part, a certain completeness condition turns out to be useful; see Chapter 13.2. It allows the description of the semi-group of invariant types up to domination-equivalence, and a characterization of forking in ACVF over very rich bases. For the most part however neither of these play any role; the significant condition is richness over the stable and the o-minimal parts. Here we adopt Kaplansky's maximally complete fields. An algebraically closed valued is maximally complete if it has no proper immmediate extensions. It follows from ([26], [27]) that any model of ACVF embeds in a maximally complete field, uniquely up to isomorphism. Since we use all the geometric sorts, a 'rich base' for us is a model of ACVF whose field part is maximally complete.

Over such a base C, we prove first, using standard results on finite dimensional vector spaces over maximally complete fields, that any field extension F is dominated by its parts in the residue field $k(F)$ and the value group Γ_F. This kind of domination does not admit descent. A stronger statement is that F is dominated by the stable part *over* C together with Γ_F, so that the type of any element of F^n over $C \cup \Gamma_F$ is stably dominated. After an algebraic interpretation of this statement, it is deduced from the previous one by a perturbation argument. Both these results are then extended to imaginary elements.

We interpret the last result as follows: an arbitrary type lies in a family of stably dominated types, definably indexed by Γ. Note that k and Γ play asymmetric roles here. Indeed, at first approximation, we develop what can be thought of as the model theory of k^Γ, rather than $k \times \Gamma$. However k^Γ is presented by a Γ-indexed system of *opaque* equivalence relations, each hiding the structure on the finer ones until a specific class is chosen. This kind of phenomenon, with hidden forms of k^n given by finitely many nested equivalence relations, is familiar from stability theory; the presence of a definable directed system of levels is new here.

Even for fields, the stable domination in the stronger statement cannot be understood without imaginaries. Consider a field extension F of C; for simplicity suppose the value group Γ_F of F is generated over Γ_C by one element γ. There is then a canonical vector space V_γ over the residue field. If γ is viewed as a code for a closed ball $E_\gamma = \{x : v(x) \geq \gamma\}$, the elements of V_γ can be taken to be codes for the maximal open sub-balls of E_γ. The vector space V_γ lies in the stable part of the theory, over $C(\gamma)$. We show that F is dominated over $C(\gamma)$ by elements of $k(F) \cup V_\gamma(F)$. Note that $k(F)$ may well be empty.

Over arbitrary bases, invariant types orthogonal to the value group are shown to be dominated by their stable part; this follows from existence of invariant extensions, and descent.

At this point, we have the metastability of ACVF. We now seek to relate this still somewhat abstract picture more directly with the geometry of valued fields. We characterize the stably dominated types as those invariant types that are orthogonal to the value group (Chapter 8.) In Chapter 14, we describe geometrically the connection between a stably dominated type P and the associated invariant type p, when P is contained in an algebraic variety V. In the case of ACF, the invariant extension is obtained by avoiding all proper subvarieties. In ACVF, the demand is not only to avoid but to stay as far away as possible from any given subvariety. See Theorem 14.12. In ACF the same prescription yields the unique invariant type of any definable set; it is not necessary to pass through types. In ACVF the picture for general definable sets is more complicated. But for a definable subgroup G of $\mathrm{GL}_n(K)$, or for a definable affine homogeneous space, we show that a translation invariant stably dominated type is unique if it exists, and that in this case it is again the type of elements of maximal distance from any proper subvariety of the Zariski closure of G.

In chapter 15 the ideas are similar, but the focus is on canonical bases. Any definable type, in general, has a smallest substructure over which it is defined. In ACF, this is essentially the field of definition of the associated prime ideal. We obtain a similar geometric description for stably dominated types; the ideal of regular functions vanishing on the type is replaced by the R-module of functions taking small values on it.

While presented here for *stably dominated types*, where the theory flows smoothly from the main ideas, within the text we try to work with weaker hypotheses on the types when possible. Over sufficiently rich base structures, all our results can be read off from the main domination results discussed above. But over smaller bases this is not always the case, leading us to think that perhaps a general principle remains to be discovered. An example is the theorem of Chapter 10, that an indiscernible sequence whose canonical base (in an appropriate sense) is orthogonal to Γ, is in fact an indiscernible set, and indeed a Morley sequence for a stably dominated type. Others are phrased in the language of *independence of substructures*.

Classical stability theory yields a notion of independence of two substructures A, B over their intersection C, defined in many equivalent ways. One is directly connected to invariant types: if A is generated by elements a, then A, B are independent over C iff $\mathrm{tp}(a/B)$ has a C-invariant extension to any model. Intuitively 'A is independent from B over C' should say that 'B provides as little as possible extra information about A, beyond what C provides'. In other words, the locus of A over B is as large as possible inside the locus of A over C. A number of the above ideas lead to notions of independence for substructures of models of ACVF, i.e., for valued fields.

The simplest notion, *sequential independence* (Chapter 8), depends on the choice of an ordered tuple a of generators of A over C. Let p be the invariant extension of $\mathrm{tp}(a/C)$ constructed above by dévissage. We say that A is sequentially independent from B over C, $A \underset{C}{\overset{g}{\downarrow}} B$, if $\mathrm{tp}(a/B) \subset p$. In general, the notion depends on the order of the tuple, and is not symmetric.

A point in an irreducible variety is generic if it does not lie in any smaller dimensional variety over the same parameters, and in an algebraically closed field, A is independent from B over C if every tuple from A which is generic in a variety defined over C remains generic in the same variety with the additional parameters from $C \cup B$. Since varieties are defined by polynomial equations, this is equivalent to saying that for every $a \in A$, the ideal of polynomials which vanish on the $\mathrm{tp}(a/C)$ is the same as those which vanish on $\mathrm{tp}(a/C \cup B)$. We extend both of these points of view to an algebraically closed valued field.

In this setting, the definable sets depend on the valuation as well, so we consider the set of polynomials which satisfy a valuation inequality on $\mathrm{tp}(a/C)$. This is no longer an ideal, but naturally gives a collection of modules over the valuation ring, which we call $J(\mathrm{tp}(a/C))$. We define A to be J-independent from B over C if $J(\mathrm{tp}(a/C)) = J(\mathrm{tp}(a/C \cup B))$ for all tuples a from A (Chapter 15). This definition does not depend on the order of the tuple a.

Our final notion of independence is defined here only for variables of the field sort. We define $\mathrm{tp}^+(A/C)$ to be the positive quantifier-free type of A over C, and say that A and B are *modulus independent* over C if $\mathrm{tp}^+(AB/C)$ is determined by the full types of A and B separately (Chapter 14). In a pure algebraically closed field, the positive type corresponds precisely to the ideal of polynomials which vanish on the type, so modulus independence is in that setting another way of stating non-forking. In an algebraically closed valued field, we use modulus independence as a step from sequential independence to J-independence. In this setting, the quantifier-free positive formulas refer to the maximum norm that a polynomial can take on a type.

All these notions agree on stably dominated types, and have the good properties of independence for stable theories (see Theorem 15.9). Under more general conditions, they diverge, and various properties can fail; for instance, for non-stably dominated types, J-independence need not be symmetric in A and B, nor need tp(A/C) have a J-independent extension over $C \cup B$. We give numerous examples showing this. We do show however that if A and B are fields, and $C \cap K \leq A$ with $\Gamma(C) = \Gamma(A)$, then sequential independence over C implies both modulus independence and J-independence.

In the final chapter we briefly illustrate the idea mentioned in the preface, that the methods of this monograph should be useful for valued fields beyond ACVF. Theorem 16.7 asserts that Scanlon's model completion of valued differential fields is metastable. This gives for the first time a language to pose structural questions about this rich theory, and we hope it will be fruitful. We also show the metastability of the theory of $\mathbb{C}((t))$, and related theories. Here we prove nothing anew, reducing all questions to properties of ACVF. The property of metastability itself can only hold for a limited number of theories of valued fields, but the method of reduction to the algebraic closure is much more general.

PART 1

STABLE DOMINATION

CHAPTER 2

SOME BACKGROUND ON STABILITY THEORY

We give here a brief preview of stability theory, as it underpins stable domination. We also introduce some of the model-theoretic notation used later. Familiarity with the basic notions of logic (languages, formulas, structures, theories, types, compactness) is assumed, but we explain the model theoretic notions beginning with saturation, algebraic closure, imaginaries. We have in mind a reader who is familiar with o-minimality or some model theory of valued fields, but has not worked with notions from stability. Sources include Shelah's *Classification Theory* as well as books by Baldwin [4], Buechler [7], Pillay [40] and Poizat [43]. There is also a broader introduction by Hart [11] intended partly for non-model theorists, and an introduction to stability theory intended for a wider audience in [18] . Most of the stability theoretic results below should be attributed to Shelah. Our treatment will mostly follow Pillay [40].

Stability theory is a large body of abstract model theory developed in the 1970s and 1980s by Shelah and others, but having its roots in Morley's 1965 Categoricity Theorem: if a complete theory in a countable language is categorical in some uncountable power, then it is categorical in all uncountable powers. Shelah formulated a radical generalization of Morley's theorem, weakening the categoricity assumption from one isomorphism type to any number less than the set-theoretic maximum. The conclusion is that all models of the theory, in any power, are classifiable by a small tree of numerical dimensions. This can be viewed as a description and analysis of all complete theories in which the large models are classifiable by a small family of numerical invariants. This work brought out an impressive list of model-theoretic ideas and properties beyond stability itself: superstability, regular types, domination and orthogonality, higher properties such as shallowness, NDOP and NOTOP. These yield sharp tools for taking structures apart, explaining much in terms of simpler embedded structures, the regular types. These achievements are analogous to Ax–Kochen, Ershov theorems in valued fields, where many properties of Henselian valued fields (with appropriate additional assumptions) are reduced to the residue field and value group. However, in stability no special parts of the theory are given in advance. All notions are defined in terms of an abstractly defined notion of independence.

Among fundamental mathematical theories, a few are stable: algebraically closed fields (ACF), and more generally, separably closed fields (SCF); differentially closed fields (the model companion of the theory of fields with a differential operator, with theory denoted DCF); and modules over any ring. By recent work of Sela, the theory of non-abelian free groups is also stable.

Since the mid 1980s, model-theorists have noticed stability-theoretic phenomena in a number of structures which formally are unstable but have significant mathematical interest. These include certain structures with *simple theory*, where the notion of independence is slightly less constrained: for example, smoothly approximable structures, pseudofinite fields, algebraically closed fields with a generic automorphism (ACFA).

Many structures of algebraic geometry can be interpreted without quantifiers in fields or in differential or difference fields. The stability or simplicity of ACF and SCF, DCF, and ACFA thus opens the way for the use of stability theoretic methods in algebraic geometry. However many other algebraic geometric and number theoretic constructions involve valuations, and these were not accessible up to now.

In a different direction, a totally ordered structure $(M, <, \dots)$ is *o-minimal* if every parameter-definable subset of M is a finite union of singletons and open intervals. Because of the total ordering, any o-minimal theory is unstable. Ideas from stability theory have influenced the development of o-minimality, but for the most part o-minimal technique uses topologies and ideas closer to classical geometric ones, especially in the local (or definably compact) parts of the theory. In Part II, we make occasional reference to o-minimality, since the value group of an algebraically closed valued field is o-minimal, but essentially no knowledge of o-minimality is needed.

The theory of stable domination, which we develop in Part I, provides a new setting where stability theory has application, for unstable structures with a stable constituent. The motivating example, as developed in Part II of this monograph, is the theory of algebraically closed valued fields.

The stability-theoretic features of ACVF as developed in Part II are seen most strongly through stable domination. However, we also prove several other results which have analogues for stable theories. For example, in Theorem 8.27 we show that in ACVF, over an algebraically closed set definable types are dense in the Stone space (in a stable theory, all types are definable). In Corollary 8.24, we show that in ACVF strong type and Lascar strong type agree. And in Proposition 10.16, we show that any indiscernible sequence over the value group is indiscernible as a *set* over the value group (in a stable theory, any indiscernible sequence is an indiscernible set). Furthermore, we develop

several different theories of independence, each of which has some properties of independence defined through non-forking in a stable theory. Below
we summarise some basic stability-theoretic facts and terminology used later,
emphasising stability ideas which lift to ACVF. Proofs are omitted, and the
presentation is mostly taken from Pillay [40].

2.1. Saturation, the universal domain, imaginaries

For this chapter, we assume that $\mathcal{L}$ is a first order language, and that T is
a complete theory over $\mathcal{L}$. Our convention will be to use a single variable x
(rather than $\bar{x}$) for a finite sequence of variables, and a single parameter a
for a finite sequence of parameters. We often omit union signs, writing for
example $ABcd$ for $A \cup B \cup \{c, d\}$, where A, B are sets of parameters. Often
we write $a \equiv_C b$ to mean that $\text{tp}(a/C) = \text{tp}(b/C)$. We shall use symbols
A, B, C for sets of parameters, and M, N for models of an ambient theory
(usually denoted T) which is understood from the context. When we say
that, for some model M, the set $X \subseteq M^n$ is *definable*, we mean that it is
definable with parameters, i.e., that it is the solution set in M^n of some formula
$\varphi(x_1, \ldots, x_n, a_1, \ldots, a_m)$ where $a_1, \ldots, a_m \in M$.

Recall that if $C \subseteq M \models T$ then the *algebraic closure of* C, written $\text{acl}(C)$
consists of those $c \in M$ such that, for some formula $\varphi(x)$ over C, $\varphi(M)$
is finite and contains c. The *definable closure* $\text{dcl}(C)$ is the union of the 1-
element C-definable sets. Thus, $C \subseteq \text{dcl}(C) \subseteq \text{acl}(C)$, $\text{dcl}(C)$ is definably
closed (that is, $\text{dcl}(\text{dcl}(C)) = \text{dcl}(C)$), and $\text{acl}(C)$ is algebraically closed.

In general, the language $\mathcal{L}$ is assumed to be multi-sorted. By 'type' we shall
always mean 'complete type over some small base set', and if completeness is
not assumed we say 'partial type'. For any set C of parameters in a model
of T, and any set x of variables, we write $S_x(C)$ for the set of types over C in
the variables x, and $S(C)$ when the variables are understood from the context.
This is more correct for many-sorted theories, since the variables in x carry
with them a specification of the relevant sort. The space $S_x(C)$ is the Stone
space of the Boolean algebra of formulas in free variables x up to equivalence
over T, and is a compact totally disconnected topological space. When we
have a particular sort in mind, we write $S_n(C)$ for $S_{x_1, \ldots, x_n}(C)$, with x_i of that
sort. (Note that the set of n-types of all sorts combined is not compact.)
The basic open sets have the form $[\varphi] = \{p \in S_x(C) : \varphi \in p\}$, where φ is a
formula in free variables among x over C. In particular, an *isolated type* is a
type which contains a formula $\varphi(x)$ (an *isolating formula*) such that for all a
(in an ambient model of T), if $\varphi(a)$ holds, then a realises p.

The set of variables x is usually assumed to be finite, but for general questions this is not needed. Types in infinitely many variables are called *-types
in [49], and they are often convenient to use. In particular, in this monograph

we often talk of tp(A/C), where A, C are infinite sets. Implicitly, we have in mind some enumeration $(a_i : i < \lambda)$ of A, and are considering a type in the variables $(x_i : i < \lambda)$, consisting of all formulas $\varphi(x_{i_1}, \ldots, x_{i_n})$ (for any $n < \omega$) over A such that $\varphi(a_{i_1}, \ldots, a_{i_n})$ holds. Alternatively, choose a variable x_a of the appropriate sort for each $a \in A$, and use the tautological correspondence of variable with element. This makes sense as long as one discusses properties of types that are invariant under permutations of the variables.

Following Morley, we present various model-theoretic ranks in terms of Cantor–Bendixson ranks of type spaces. Recall that if X is a compact totally disconnected topological space then the Cantor–Bendixson rank $\mathrm{CB}_X(p)$ of $p \in X$ is defined by transfinite induction as follows: $\mathrm{CB}_X(p) \geq 0$ for all $p \in X$, and $\mathrm{CB}_X(p) = \alpha$ if p is isolated in the closed subspace $\{q \in X : \mathrm{CB}_X(q) \geq \alpha\}$. If $\mathrm{CB}_X(p) < \infty$ for all $p \in X$, then by topological compactness $\{\mathrm{CB}_X(p) : p \in X\}$ has a maximal element α, and $\{p \in X : \mathrm{CB}_X(p) = \alpha\}$ is finite; its size is denoted by $\mathrm{CB\text{-}Mult}(X)$.

If λ is an infinite cardinal, the model $M \models T$ is λ-saturated, if, for all $C \subset M$ with $|C| < \lambda$, M realises all types in $S(C)$; it is saturated if it is $|M|$-saturated. We shall work in a universal domain $\mathcal{U}$, which is assumed to be a large model of T. Any parameter sets $A, B, C, \ldots$ which we mention are subsets of $\mathcal{U}$, and any models $M, N, \ldots \models T$ are assumed to be elementary substructures of $\mathcal{U}$. We write $\varphi(a)$ to mean: $\mathcal{U} \models \varphi(a)$. We often consider types $p \in S(\mathcal{U})$, and occasionally realisations a of p, which in general are not in $\mathcal{U}$. It is common to assume that $\mathcal{U}$ is saturated, since this guarantees that any elementary map between subsets of $\mathcal{U}$ of size less than $|\mathcal{U}|$ extends to an element of $\mathrm{Aut}(\mathcal{U})$. For unstable theories (such as ACVF), existence of saturated models depends on set-theoretic assumptions, but a variety of set-theoretic tools (such as absoluteness) can be used to dispense with these a posteriori. For most arguments it suffices to assume that $\mathcal{U}$ is sufficiently saturated. Some cardinal κ is fixed, and sets of size $< \kappa$ are called small; on the other hand $\mathcal{U}$ is assumed to be κ-saturated and κ-homogeneous, i.e., two elements realizing the same type over a small set A are conjugate under the automorphism groups. These two conditions can be achieved without any special set-theoretic assumptions or tools; κ is chosen safely above the cardinalities of any objects of interest.

If T is a multi-sorted theory, S is a sort of T, and $M \models T$, write $S(M)$ for the set of elements of M of sort S. We say that T has elimination of imaginaries if, for any $M \models T$, any collection $S_1, \ldots, S_k$ of sorts in T, and any $\emptyset$-definable equivalence relation E on $S_1(M) \times \cdots \times S_k(M)$, there is a $\emptyset$-definable function f from $S_1(M) \times \cdots \times S_k(M)$ into a product of sorts of M, such that for any $a, b \in S_1(M) \times \cdots \times S_k(M)$, we have Eab if and only if $f(a) = f(b)$. Given a complete theory T, it is possible to extend it to a complete theory T^{eq} over a language L^{eq} by adjoining, for each collection

$S_1, \ldots, S_k$ of sorts and $\emptyset$-definable equivalence relation E on $S_1 \times \cdots \times S_k$, a sort $(S_1 \times \cdots \times S_k)/E$, together with a function symbol for the natural map $a \mapsto a/E$. Any $M \models T$ can be canonically extended to a model of T^{eq}, denoted M^{eq}. The theory T^{eq} automatically has elimination of imaginaries. For the purposes of stability theory, elimination of imaginaries is very helpful. Therefore, in the development of stability theory in this chapter we shall assume that T has elimination of imaginaries (though not necessarily that T is formally a theory of form $(T')^{\mathrm{eq}}$). We shall refer to the sorts of T^{eq} as imaginary sorts, and to elements of them as *imaginaries*.

Suppose that D is a definable set in $M \models T$, defined say by the formula $\varphi(x, a)$. There is a $\emptyset$-definable equivalence relation $E_\varphi(y_1, y_2)$, where $E_\varphi(y_1, y_2)$ holds if and only if $\forall x(\varphi(x, y_1) \leftrightarrow \varphi(x, y_2))$. Now a/E_φ is identifiable with an element of an imaginary sort; it is determined uniquely (up to interdefinability over $\emptyset$) by D, and will often be referred to as a *code* for D, and denoted $\ulcorner D \urcorner$. We prefer to think of $\ulcorner D \urcorner$ as a fixed object (e.g., as a member of $\mathcal{U}^{\mathrm{eq}}$) rather than as an equivalence class of M; for viewed as an equivalence class it is formally a different set (as is D itself) in elementary extensions of M.

Under our assumption that T eliminates imaginaries, $\ulcorner D \urcorner$ can be regarded as a tuple in M; without elimination of imaginaries we would just know it to be in M^{eq}. An automorphism of M will fix D setwise if and only if it fixes $\ulcorner D \urcorner$.

In a purely theoretical context, the device of moving to M^{eq} gives a soft way of dealing with imaginaries. However when working with specific theories a more concrete description of a collection of sorts admitting elimination of imaginaries is useful. It turns out that o-minimal fields, separably and differentially closed fields, and existentially closed difference fields all admit elimination of imaginaries in the field sort. For algebraically closed valued fields, the field sort, value group and residue field do not exhaust the necessary imaginaries. The main purpose of [12] was to identify a geometrically meaningful family of sorts which, when adjoined to the field sort, suffice for elimination of imaginaries. See Chapter 7 for a description.

2.2. Invariant types

Let $\mathrm{Aut}(\mathcal{U}/C)$ denote the subgroup of $\mathrm{Aut}(\mathcal{U})$ fixing C pointwise. Then $\mathrm{Aut}(\mathcal{U}/C)$ acts naturally on $S(\mathcal{U})$: if $g \in \mathrm{Aut}(\mathcal{U}/C)$ then $g(p) = \{\varphi(x, g(a)) : \varphi(x, a) \in p\}$.

DEFINITION 2.1. The type $q \in S(\mathcal{U})$ is $\mathrm{Aut}(\mathcal{U}/C)$-*invariant* if it is fixed by this action.

If q is $\mathrm{Aut}(\mathcal{U}/C)$-invariant, then for any formula $\varphi(x, y)$ and $a_1, a_2 \in \mathcal{U}$, if $a_1 \equiv_C a_2$ then $\varphi(x, a_1) \in q$ if and only if $\varphi(x, a_2) \in q$, and assuming saturation this statement is equivalent to invariance. This says that $p \in S(\mathcal{U})$ is Aut($\mathcal{U}/C$)-invariant precisely if it does not *split* over C, in the sense of Shelah [49]. This gives an alternative definition of *invariant type*, without reference to automorphisms and saturation assumptions.

Suppose that $p \in S(\mathcal{U})$ is Aut($\mathcal{U}/C$)-invariant, and $e_1, e_2 \in \mathcal{U}$ with $e_1 \equiv_C e_2$. Let $a \models p|Ce_1e_2$. Then $e_1 \equiv_{Ca} e_2$. This observation will be used frequently without explicit mention.

Call a type p over $\mathcal{U}$ *invariant* if it is Aut($\mathcal{U}/A$) -invariant for some small A (in the sense of the previous section.) Any such A is called a *base* for p.

Let Inv_x denote the set of invariant types in the variable x. If $p \in \mathrm{Inv}_x$, $q \in \mathrm{Inv}_y$ are invariant types, define $p \otimes q$ as follows: let A be small such that p is Aut($\mathcal{U}/A$) -invariant; let $d \models q|A$, and let $c \models p|Ad$. Then $\mathrm{tp}(cd/A)$ does not depend on the choice of c, d; call it $(p \otimes q)|A$. Clearly there exists a unique type $p \otimes q$ over $\mathcal{U}$ whose restriction to any such A is $(p \otimes q)|A$. Moreover $p \otimes q$ is Aut($\mathcal{U}/B$)-invariant whenever p, q are. Thus $p \otimes q \in \mathrm{Inv}_{xy}$. This gives an associative product, not in general commutative.

We take this opportunity to observe that a notion of *domination* can be defined for invariant types in full generality. In stable theories, it will agree with the usual notion. Write $P \equiv Q$ if the two partial types P, Q imply each other.

DEFINITION 2.2. If p, q are invariant types over $\mathcal{U}$, call p, q *domination-equivalent* if for some base A for p, q and some $r \in S_{xy}(A)$ containing $p(x)|A \cup q(y)|A$, we have $p(x) \cup r(xy) \equiv q(y) \cup r(xy)$

It is easy to see that domination-equivalence is an equivalence relation on invariant types, and is a congruence with respect to $\otimes$. The set of domination-equivalence classes of invariant types thus becomes an associative, commutative semigroup. Denote it by $\overline{\mathrm{Inv}}(\mathcal{U})$.

We will use domination-equivalence almost exclusively when one of these types lies in the stable part of the theory; in this case the other will be said to be *stably dominated*.

2.3. Conditions equivalent to stability

We give some definitions which yield notions equivalent to stability.

DEFINITION 2.3. (i) The theory T is λ-*stable* (where λ is an infinite cardinal), if for all $C \subset \mathcal{U}$ with $|C| = \lambda$, $|S(C)| = \lambda$. It is *stable* if it is λ-stable for some infinite λ.

(ii) The formula $\varphi(x, y)$ (possibly with extra parameters) is *unstable* if there are $a_i, b_i \in \mathcal{U}$ for $i < \omega$ such that for all $i, j < \omega$, $\varphi(a_i, b_j)$ holds if and only if $i < j$. We say $\varphi(x, y)$ is *stable* if it is not unstable.

(iii) Let $p(x) \in S(B)$ and $C \subseteq B$. Then p is *definable over* C if for every $\mathcal{L}$-formula $\varphi(x, y)$ there is an $\mathcal{L}(C)$-formula $\psi(y)$ (often denoted $(d_p x)(\varphi(x, y))$) such that, for all $b \in B$, $\varphi(x, b) \in p$ if and only if $\psi(b)$ holds. We say that p is definable *almost over* C if it is definable over $\mathrm{acl}(C)$ *imaginaries included*, and that p is *definable* if it is definable over B. An isolated type p over B (isolated via a formula $\psi(x)$) is an example of a definable type over B in this sense, since $\varphi(x, b) \in p$ if and only if $(\forall x)(\psi(x) \rightarrow \varphi(x, b))$. However, this definition does not extend to structures B' containing B. Definable types are useful inasmuch as they are 'extendible', i.e., the same definition scheme defines a complete type $p|B'$ over any bigger base set B'. In this monograph, we will only consider definable types which are extendible, and indeed, think of the function $B' \mapsto (p|B')$ as being the definable type. Definable types over models are always uniquely extendible; much less trivially, in stable theories with elimination of imaginaries, definable types over algebraically closed sets are uniquely extendible. The uniqueness is the content of the *finite equivalence relation theorem*, mentioned below.

(iv) Suppose that Δ is a finite set of formulas $\delta(x, y)$, and let $S_\Delta(\mathcal{U})$ denote the space of complete Δ-types over $\mathcal{U}$; here a Δ-type is a maximal (subject to consistency with T) set of Δ-formulas, that is, Boolean combinations of formulas $\delta(x, a)$ for $\delta \in \Delta$ and $a \in \mathcal{U}$. Then $S_\Delta(\mathcal{U})$ is a compact totally disconnected topological space, where a basic open set is the collection of Δ-types containing a fixed Δ-formula. Let $\Phi(x)$ be a set of formulas over some $A \subset \mathcal{U}$ (so $|\Phi| < |\mathcal{U}|$). Then the Δ-*rank* of $\Phi(x)$, denoted $R_\Delta(\Phi(x))$, is the Cantor–Bendixson rank of the subspace $Y = \{q \in S_\Delta(\mathcal{U}): q \text{ is consistent with } \Phi(x)\}$ of $S_\Delta(\mathcal{U})$. If this is finite, then $\mathrm{Mult}_\Delta(\Phi)$ is the Cantor–Bendixson multiplicity of Φ.

THEOREM 2.4. *The following are equivalent.*

(i) *T is stable.*
(ii) *Every formula $\varphi(x, y)$ is stable.*
(iii) *For every model M, every 1-type over M is definable.*
(iv) *For every model M, every type over M is definable.*
(v) *For every Δ and Φ as in Definition 2.3, $R_\Delta(\Phi)$ is finite.*
(vi) *For every Δ, the space $S_\Delta(M)$ has cardinality at most $|M| + |L|$.*
(vii) *For some κ, for any M, the space of types $S(M)$ over M has cardinality at most $|M|^\kappa$.*

(vii) is often the easiest way to verify stability, while (v) is most useful in proofs.

REMARK 2.5. Suppose that $\varphi(x, y)$ is a stable formula, $C \subset M$ with M $|C|^+$-saturated, and $p \in S(M)$. Then there are $a_1, \ldots, a_n \in M$ with $a_i \models p|C \cup \{a_j : j < i\}$ for each $i = 1, \ldots, n$, such that $(d_p x)\varphi(x, y)$ is equivalent to a *positive* Boolean combination of formulas $\varphi(a_i, y)$. See for example Lemma 1.2.2 of [40] for a proof.

If $C \subseteq M \models T$ and $p \in S(M)$ is C-definable, then the *defining schema* for p (the map $\varphi(x, y) \mapsto (d_p x)\varphi(x, y))$ yields canonically an extension $p|B$ of p over any $B \supseteq C$, and $p|B$ is also C-definable. In particular, there is a canonical extension $p|\mathcal{U}$ to $\mathcal{U}$. In the other direction, to restrict the parameter set, if q is *any* type over $\mathcal{U}$, then $q|C$ denotes the restriction of q to C, so $q|C \in S(C)$.

Clearly, any C-definable type over $\mathcal{U}$ is $\mathrm{Aut}(\mathcal{U}/C)$-invariant, but other (undefinable) examples of invariant types play a major role in Part II.

2.4. Independence and forking

Stable theories admit a unique theory of independence, that can be approached in a number of ways. One such approach, that we will not mention, is the Lascar–Poizat notion of *fundamental order*. Another, essential in the generalization to simplicity, de-emphasizes uniqueness properties in favour of their consequences in terms of amalgamation of independent triangles. We will focus on two aspects: (1) the choice of a canonical element of a type space; especially, if $C \subseteq B \subset \mathcal{U}$ and $p \in S(C)$, a canonically chosen type $q \in S(B)$ which extends p. (2) For the best such q, the information content in q should be as small as possible, and the locus of q (i.e., the set of realisations in $\mathcal{U}$) will in some sense be as large as possible. On the other hand, the other extensions will be 'small'; a reasonable technical notion of smallness is given by 'dividing', below.

DEFINITION 2.6. Let $C \subseteq B \subset \mathcal{U}$ and $\varphi(x, b) \in S(B)$ (so $b \in B$). Then $\varphi(x, b)$ *divides* over C if there is $k \in \omega$ and a sequence $(b_i : i \in \omega)$ of realisations of $\mathrm{tp}(b/C)$ such that $\{\varphi(x, b_i) : i \in \omega\}$ is k-inconsistent, that is, any subset of it of size k is inconsistent. A partial type $\pi(x)$ over B *forks* over C if there are $n \in \omega$ and formulas $\varphi_0(x), \ldots, \varphi_n(x)$ such that $\pi(x)$ implies $\bigvee_{i \leq n} \varphi_i(x)$, and each $\varphi_i(x)$ divides over C. If $p \in S(B)$ does not fork over C and a realises p, we write $a \underset{C}{\smile} B$, and say that a is independent from B over C. More generally, if A, B, C are subsets of $\mathcal{U}$, we write $A \underset{C}{\smile} B$ if for any finite tuple a from A, $a \underset{C}{\smile} B \cup C$.

THEOREM 2.7. *Let T be stable, and $C \subseteq B$. Then*

(i)   *if $p \in S(C)$, then p has a non-forking extension over B,*
(ii) *if $q \in S(B)$, then q does not fork over C if and only if it is definable almost over C,*
(iii) *if $q \in S(B)$, then q does not fork over C if and only if $R_\Delta(q|C) = R_\Delta(q)$ for all finite sets $\Delta(x)$ of $\mathcal{L}$-formulas.*

A *finite equivalence relation* is a definable equivalence relation with finitely many classes. By the following theorem, in a stable theory the non-forking extensions of a type are governed by finite equivalence relations.

THEOREM 2.8 (Finite Equivalence Relation Theorem). *Assume T is stable. Let $C \subseteq B \subset \mathcal{U}$, and $p_1, p_2 \in S(B)$ be distinct non-forking extensions of $p \in S(C)$. Then there is a C-definable finite equivalence relation E such that $p_1(x) \cup p_2(y) \vdash \neg Exy$.*

Under our assumption of elimination of imaginaries, the statement becomes simpler: let $C = \text{acl}(C) \subseteq B \subset \mathcal{U}$, and let $p_1, p_2 \in S(B)$ be non-forking extensions of $p \in S(C)$. Then $p_1 = p_2$.

A type $p \in S(C)$ is *stationary* if it has a unique non-forking extension over $\mathcal{U}$. By the Finite Equivalence Relation Theorem, in a stable theory, if $C = \text{acl}(C) \subset \mathcal{U}$, then every type over C is stationary: indeed, by elimination of imaginaries, the equivalence classes of the equivalence relation given in Theorem 2.8 are coded by an element of $\mathcal{U}$, and this element will lie in $\text{acl}(C)$. In particular, any type over a model is stationary. If $p \in S(C)$ is stationary, and $C \subset B$, we write $p|B$ for the unique non-forking extension of p to B.

If $C \subset \mathcal{U}$, we say that a, b have the same *strong type* over C, written, $\text{stp}(a/C) = \text{stp}(b/C)$, if a, b lie in the same class of any C-definable finite equivalence relation. Equivalently, with imaginaries: $\text{tp}(b/\text{acl}(C)) = \text{tp}(a/\text{acl}(C))$.

For the sake of one result (Corollary 8.24), we mention the following variant, which has been important in the study of simple theories: a and b have the same *Lascar strong type* over C if there is a sequence $M_1, \ldots, M_n$ of models containing C, and for each $i = 1, \ldots, n$ some $f_i \in \text{Aut}(\mathcal{U}/M_i)$, such that $b = f_n \circ \cdots \circ f_1(a)$. In stable theories, the notion of strong type and Lascar strong type coincide, but there are other (so far, all artificially constructed) theories in which they are known to differ. A major question is whether they agree in all simple theories.

REMARK 2.9. Suppose $C \subseteq B$.
(i) If $\text{tp}(a/B)$ has an $\text{Aut}(\mathcal{U}/\text{acl}(C))$-invariant extension over $\mathcal{U}$, then $a \underset{C}{\downarrow} B$.
(ii) If T is stable and $a \underset{C}{\downarrow} B$, then $\text{tp}(a/B)$ has a unique $\text{Aut}(\mathcal{U}/\text{acl}(C)$-invariant extension over $\mathcal{U}$.

Here, (i) follows almost immediately from the definition of forking, and (ii) is a consequence of 2.8. Under our ambient assumption of elimination of imaginaries, we can work with $\text{acl}(C)$, not $\text{acl}^{\text{eq}}(C)$.

Here is a standard list of basic properties of the non-forking relation $A \underset{C}{\downarrow} B$ on sets in a stable theory.
(i) (Existence) If $p \in S(C)$ and M is a model containing C, then p has a non-forking extension $q \in S(M)$; if $C = \text{acl}(C)$ then q is unique.

(ii) (Finite character) If $p \in S(B)$ and $B \supseteq C$, then p does not fork over C if and only if for every formula $\varphi(x) \in p$, φ does not fork over C. In particular, p does not fork over C if and only if for all finite $B_0 \subseteq B$, $p|CB_0$ does not fork over C.

(iii) (Transitivity) If $p \in S(B)$ and $C \subseteq D \subseteq B$ then p does not fork over C if and only if p does not fork over D and $p|D$ does not fork over C.

(iv) (Symmetry) If $C \subseteq B$, then $\operatorname{tp}(a/B)$ does not fork over C if and only if for all finite tuples b from B, $\operatorname{tp}(b/Ca)$ does not fork over C.

(v) (Invariance) If $p \in S(B)$ does not fork over $C \subset B$, and $f \in \operatorname{Aut}(\mathcal{U})$, then $f(p) \in S(f(B))$ does not fork over $f(C)$.

(vi) If $p \in S(C)$ and $B \supseteq C$ then p has at most $2^{|T|}$ non-forking extensions over B.

(vii) If $p \in S(B)$ then there is $C \subseteq B$ with $|C| \leq |T|$ such that p does not fork over C.

(viii) If $p \in S(B)$ has finitely many realisations and does not fork over $C \subset B$, then $p|C$ has finitely many realisations.

REMARK 2.10. Symmetry of non-forking can also be viewed locally. Let $\delta(x, y)$ be a stable formula, and let $p(x) \in S_\delta(\mathcal{U})$, $q(y) \in S_\delta(\mathcal{U})$. These are both definable – this comes from the proof of Theorem 2.4. Let $\psi(y)$ be the δ-definition $d_p x \delta(x, y)$ of p, and $\chi(x)$ be the δ-definition $d_q y \delta(x, y)$ of q. Then $\chi(x) \in p(x)$ if and only if $\psi(y) \in q(y)$ – see [40, Lemma I.2.8].

In Part I we make occasional use of the notion of *weight*. In a stable theory, T, the *preweight* of a type $p(x) = \operatorname{tp}(a/C)$ is the supremum of the set of cardinals κ for which there is an C-independent set $\{b_i : i < \kappa\}$ such that $a \downarrow_C b_i$ for all i. The *weight* $\operatorname{wt}(p)$ is the supremum of $\{\operatorname{prwt}(q) : q$ is a non-forking extension of $p\}$.

LEMMA 2.11. *In a stable theory, any type has weight bounded by the cardinality of the language; in a superstable theory, any type in finitely many variables has finite weight.*

PROOF. See Pillay [40] ch. 1, Section 4.4 or Section 5.6.3 of Buechler [7]. □

Suppose that the theory T is stable. Let $p \in S_x(C)$ be stationary, and $\varphi(x, y)$ be a formula. By definability of p, there is a φ-definition of p, namely, a formula $\psi(y)$ of form $(d_p x)(\varphi(x, y))$. The set defined by $\psi(y)$ has a code c, unique up to definable closure, and $c \in \mathcal{U}$ by elimination of imaginaries. The *canonical base* $\operatorname{Cb}(p)$ of p is the definable closure of the set of all codes c for sets defined by formulas $(d_p x)(\varphi(x, y))$ (x fixed, y and φ varying). If $p = \operatorname{tp}(a/C)$, we write $\operatorname{Cb}(a/C)$ for $\operatorname{Cb}(p)$. This is the smallest set over which p is defined.

LEMMA 2.12. *Assume T is stable, let $C \subseteq B$, and let $p \in S(B)$ be stationary. Then*

(i) $\operatorname{Cb}(p) \subseteq \operatorname{dcl}(B)$;

(ii) $\mathrm{Cb}(p) \subseteq \mathrm{acl}(C)$ *if and only if p does not fork over C;*
(iii) $\mathrm{Cb}(p) \subseteq \mathrm{dcl}(C)$ *if and only if p does not fork over C and $p|C$ is stationary.*

2.5. Totally transcendental theories and Morley rank

The theory T is *totally transcendental* if the Cantor–Bendixson rank of each type space $S_x(\mathcal{U})$ is ordinal-valued. If $\mathcal{L}$ is countable, this is equivalent to T being ω-stable. Let $\Phi(x)$ be a set of formulas, and $X = \{p \in S_x(\mathcal{U}): \Phi \subseteq p\}$ (a closed subspace of $S_x(\mathcal{U})$). Then the Morley rank $\mathrm{RM}(\Phi)$ of Φ is just $\mathrm{CB}(X)$. If $\mathrm{RM}(\Phi) < \infty$ (as holds if T is totally transcendental), then the Morley degree $\mathrm{DM}(\Phi)$ of Φ is CB-Mult(X). We often write $\mathrm{RM}(a/C)$ for $\mathrm{RM}(\mathrm{tp}(a/C))$, and similarly for Morley degree. In a totally transcendental theory, forking is determined by Morley rank: if $C \subseteq B$, then $a \underset{C}{\downarrow} B$ if and only if $\mathrm{RM}(a/C) = \mathrm{RM}(a/B)$.

We give a slightly more explicit definition of Morley rank and degree for a definable set D. It is defined inductively, as follows.

(i) $\mathrm{RM}(D) \geq 0$ if and only if $D \neq \emptyset$.

(ii) For a limit ordinal α, $\mathrm{RM}(D) \geq \alpha$ if and only if, for all $\lambda < \alpha$, $\mathrm{RM}(D) \geq \lambda$.

(iii) For any ordinal α, $\mathrm{RM}(D) \geq \alpha + 1$ if and only if there are infinitely many pairwise disjoint definable subsets D_i $(i < \omega)$ of D each of Morley rank at least α.

If $\mathrm{RM}(D) \geq \alpha$ but $\mathrm{RM}(D) \not\geq \alpha+1$ we write $\mathrm{RM}(D) = \alpha$, and if $\mathrm{RM}(D) \geq \alpha$ for all ordinals α we put $\mathrm{RM}(D) = \infty$. Finally, if $\mathrm{RM}(D) = \alpha$, then $\mathrm{DM}(D)$ is the largest n such that there are n disjoint definable subsets of D each of Morley rank α.

If the definable set X in $\mathcal{U}$ has Morley rank and degree both equal to 1, then X is said to be *strongly minimal*. A 1-sorted theory T is strongly minimal if the set (in one variable) defined by $x = x$ is strongly minimal. Examples of strongly minimal theories are the theories of pure sets, vector spaces, and algebraically closed fields. In an algebraically closed valued field, the residue field is an algebraically closed field with no extra structure, so is strongly minimal. Any strongly minimal theory is $\aleph_1$-categorical, and conversely, according to Baldwin and Lachlan [5], any model of an $\aleph_1$-categorical theory is prime over a certain strongly minimal set and the parameters used to define it.

2.6. Prime models

Prime models play a central role in Shelah's classification theory. We summarise the basic facts here. In Chapter 11, a slight generalisation of the notion is developed, partly in the ACVF context.

DEFINITION 2.13. Let T be a complete theory, and $C \subseteq M \models T$.

(i) M is *prime over* C if, for every $N \models T$ with $C \subseteq N$, there is an elementary embedding $f : M \to N$ which is the identity on C.

(ii) M is *atomic over* C if for any finite tuple a of elements of M, $\mathrm{tp}(a/C)$ is isolated.

(iii) M is *minimal over* C if there is no proper elementary substructure of M which contains C.

An easy consequence of the Omitting Types Theorem is that if $\mathcal{L}$ and $S(\emptyset)$ are countable then T has an atomic model over $\emptyset$. More generally, we have the following (see e.g., [36, 4.2.10]).

THEOREM 2.14. *If T is a complete theory with infinite models over a countable language, then the following are equivalent.*

(i) *T has a prime model.*

(ii) *T has an atomic model.*

(iii) *For all n, the set of isolated n-types is dense in $S_n(\emptyset)$.*

In many algebraic theories, prime models exist and have an algebraic characterisation. For the theories of algebraically closed fields and real closed fields, the prime model over C is exactly the (field-theoretic) algebraic closure of C, respectively the real closure. For differentially closed fields it is the differential closure; uniqueness and non-minimality of the differential closure was one of the early applications of stability to this area. For p-adically closed fields a similar characterization holds when C is a field, but not when imaginary elements are allowed.

THEOREM 2.15. *Let T be ω-stable or o-minimal, and $C \subset \mathcal{U}$. Then there is a (unique up to isomorphism over C) prime model M of T over C, and M is also atomic over C.*

In the above theorem, the ω-stable case is due to Shelah, and the o-minimal case due to Pillay and Steinhorn.

2.7. Indiscernibles, Morley sequences

If $(I, <)$ is a totally ordered set, then the sequence $(a_i : i \in I)$ of distinct elements of $\mathcal{U}$ is *(order)-indiscernible over* C if, for any $n \in \omega$, and $i_1, \ldots, i_n, j_1, \ldots, j_n \in I$ with $i_1 < \cdots < i_n$ and $j_1 < \cdots < j_n$, $\mathrm{tp}(a_{i_1} \ldots a_{i_n}/A) = \mathrm{tp}(a_{j_1} \ldots a_{j_n}/C)$. We say that $\{a_i : i \in I\}$ is an *indiscernible set* over C if for any distinct $i_1, \ldots, i_n \in I$ and distinct $j_1, \ldots, j_n \in I$, $\mathrm{tp}(a_{i_1} \ldots a_{i_n}/C) = \mathrm{tp}(a_{j_1} \ldots a_{j_n}/C)$ (so the order doesn't matter).

A standard application of Ramsey's theorem, compactness, and the saturation of $\mathcal{U}$ ensures that for any ordered set $(I, <)$ which is 'small' relative to $\mathcal{U}$

and any small parameter set C, there is in $\mathcal{U}$ an indiscernible sequence over C of order type I. If the theory T is stable, any infinite indiscernible sequence is an indiscernible set.

Assume now that T is stable, and let $p(x) \in S(C)$ be stationary. A *Morley sequence* for p of length λ is a sequence $(a_i : i < \lambda)$ such that for each $i < \lambda$, a_i realises $p|C \cup \{a_j : j < i\}$. Such a sequence is independent over C and is an indiscernible set over C. Furthermore, if $C = \mathrm{acl}(C)$ any infinite indiscernible set over C which is independent over C is a Morley sequence for some strong type over C.

Below, and elsewhere, if $J \subseteq I$, we write a_J for the subsequence $(a_j : j \in J)$.

LEMMA 2.16. *Let* $\{a_i : i \in I\}$ *be an infinite indiscernible set over* C, *and suppose that for any finite disjoint* $J, J' \subset I$, $\mathrm{acl}(Ca_J) \cap \mathrm{acl}(Ca_{J'}) = \mathrm{acl}(C)$. *Then* $\{a_i : i \in I\}$ *is a Morley sequence over* C.

PROOF. This follows easily from Lemma 5.1.17 of Buechler [7], which states that the indiscernible sequence I is a Morley sequence over $C \cup J$ for any infinite $J \subset I$. To apply this, let $i_1, \ldots, i_n \in I$ be distinct, and let J_1, J_2 be infinite disjoint subsets of $I \setminus \{i_1, \ldots, i_n\}$. Then $a_{i_1} \underset{CJ_j}{\downarrow} a_{i_2} \ldots a_{i_n}$ for $j = 1, 2$, so $\mathrm{Cb}(a_1/Ca_{i_2} \ldots a_{i_n}) \subseteq \mathrm{acl}(CJ_1 \cap CJ_2) = \mathrm{acl}(C)$. $\square$

2.8. Stably embedded sets

A C-definable set D in $\mathcal{U}$ is *stably embedded* if, for any definable set E and $r > 0$, $E \cap D^r$ is definable over $C \cup D$. If instead we worked in a small model M, and C, D were from M^{eq}, we would say that D is stably embedded if for any definable E in M^{eq} and any r, $E \cap D^r$ is definable over $C \cup D$ *uniformly* in the parameters defining E; that is, for any formula $\varphi(x, y)$ there is a formula $\psi(x, z)$ such that for all a there is a sequence d from D such that

$$\{x \in D^r : \ \models \varphi(x, a)\} = \{x \in D^r : \ \models \psi(x, d)\}.$$

For more on stably embedded sets, see the Appendix of [8]. We mention in particular that if D is a C-definable stably embedded set and $\mathcal{U}$ is saturated, then any permutation of D which is elementary over C extends to an element of $\mathrm{Aut}(\mathcal{U}/C)$. Basic examples of stably embedded sets include the field of constants in a differentially closed field, and the fixed field in a model of ACFA. By definability of types, in a stable theory all definable sets are stably embedded.

If D is C-definable, then we say that D is *stable* if the structure with domain D, when equipped with all the C-definable relations, is stable. We emphasise the distinction between saying that a definable set D is stable, and that a formula $\varphi(x, y)$ is a stable formula (as in Definition 2.3).

In this monograph, working over a set C of parameters, C-definable sets which are both stable and stably embedded play a crucial role. Such sets are in some sources just called *stable*.

LEMMA 2.17. *Let D be a C-definable set. Then the following are equivalent.*
(i) *D is stable and stably embedded.*
(ii) *Any formula $\varphi(x_1 \ldots x_n, y)$ which implies $D(x_1) \wedge \cdots \wedge D(x_n)$ is stable (viewed as a formula $\varphi(x, y)$, where $x = (x_1, \ldots, x_n)$).*
(iii) *If $\lambda > |T| + |C|$ with $\lambda = \lambda^{\aleph_0}$ and $B \supseteq C$ with $|B| = \lambda$, then there are at most λ 1-types over B realised in D.*

In particular, by (ii), if D is C-definable, and $C \subseteq B$, then D is stable and stably embedded over C if and only if it is stable and stably embedded over B, so there is no need to mention the parameter set. Hence, in the notation of the next chapter, any sort in St_C is also in $\mathrm{St}_{C'}$ for some finite $C' \subseteq C$; thus:

COROLLARY 2.18. *Let S be a sort of St_C. The language of S is obtained from a language of size $\leq |T|$ by the addition of constants.*

If D, E are definable sets in $\mathcal{U}$, then E is said to be D-*internal* if there is a finite set A of parameters such that $E \subset \mathrm{dcl}(D \cup A)$. It is immediate that any definable set which is internal to a stable and stably embedded set is stable and stably embedded. In ACVF, by Proposition 7.8, any stable and stably embedded set is internal to the residue field.

CHAPTER 3

DEFINITION AND BASIC PROPERTIES OF St_C

Given any theory T, we consider the stable, stably embedded definable sets, including imaginary sorts. The induced structure $\mathrm{St}_\emptyset$ on these sets can be viewed as the maximal stable theory interpretable without parameters in T. We define a class of types of T, called *stably dominated*, that are not necessarily stable, but have a stable part; and such that any interaction between a realization of the type and any other set B is preceded by an interaction of the stable part with B. See below for a precise definition.

We show that the theory of independence on the stable part lifts to the stably dominated types. The same is true of a slightly less well-known theory of independence based on dcl, rather than acl, and we describe this refinement too. We develop the basic properties of stably dominated types.

At this level of generality, elimination of quantifiers and of imaginaries can be, and will be assumed.

Let $\mathcal{U}$ be a sufficiently saturated model of a complete theory T with elimination of imaginaries. Throughout the monograph, A, B, C, D will be arbitrary subsets of the universe. Symbols a, b, c denote possibly infinite tuples, sometimes regarded as sets, sometimes as tuples, and indeed sometimes A, B are implicitly regarded as tuples. We occasionally write $a \subset X$ to mean that the set enumerated by a is a subset of X. We write $A \equiv_C B$ to mean that, under some enumeration of A, B which is fixed but usually not specified, $\mathrm{tp}(A/C) = \mathrm{tp}(B/C)$. By the saturation assumption, this is equivalent to A and B, in the given enumerations, being in the same orbit of $\mathrm{Aut}(\mathcal{U}/C)$ (the group of automophisms of $\mathcal{U}$ which fix C pointwise); sometimes we just say that A and B are *conjugate* over C.

DEFINITION 3.1. Let C be any small set of parameters. Write St_C for the multi-sorted structure $\langle D_i, R_j \rangle_{i \in I, j \in J}$ whose sorts D_i are the C-definable, stable, stably embedded subsets of $\mathcal{U}$. For each finite set of sorts D_i, all the C-definable relations on their union are included as $\emptyset$-definable relations R_j. For any $A \subset \mathcal{U}$, write $\mathrm{St}_C(A) := \mathrm{St}_C \cap \mathrm{dcl}(CA)$. We often write A^{st} for $\mathrm{St}_C(A)$ when the base C is unambiguous.

We begin with some basic observations about St_C.

LEMMA 3.2. *The stucture* St_C *is stable.*

PROOF. It suffices to observe that if D_1 and D_2 are C-definable stable and stably embedded sets, then so is $D_1 \cup D_2$. By Lemma 2.17, it is sufficient to show that if $C' \supseteq C$ and $|C'| = \lambda$, where $\lambda = \lambda^{\aleph_0} > |T| + |C|$, then there are at most λ 1-types over C' realised in $D_1 \cup D_2$. This is immediate, as by the same lemma there are at most λ 1-types realised in each of D_1 and D_2.   □

We shall use the symbol $\underset{}{\bigcup}$ for non-forking (for subsets of St_C) in the usual sense of stability theory; the context should indicate that this takes place in St_C. In particular, $A \underset{C}{\bigcup} B$ means that A, B are independent (over C) in St_C, unless we explicitly state that the independence is in another structure. Notice that, if $A, B \subset \mathrm{St}_C$, although we do not expect $A = \mathrm{St}_C(A)$, still $A \underset{C}{\bigcup} B$ if and only if $\mathrm{St}_C(A) \underset{C}{\bigcup} \mathrm{St}_C(B)$.

We begin with some remarks on how the structure St_C varies when the parameter set C increases. As noted after Lemma 2.17, the property of a definable set D being stable and stably embedded does not depend on the choice of defining parameters. Thus, if D is a sort of St_C and $B \supseteq C$, then D is also a sort of St_B. In general, if $C \subseteq B$ then St_B has more sorts than St_C. Notice that $\mathrm{St}_{\mathrm{acl}(C)}$ has essentially the same domain as St_C: if D_1 is a sort of $\mathrm{St}_{\mathrm{acl}(C)}$ whose conjugates over C are $D_1, \ldots, D_r$, then $D_1 \cup \cdots \cup D_r$ is a sort of St_C. Each element of $\mathrm{St}_{\mathrm{acl}(C)}$ has a code over C in St_C.

LEMMA 3.3. *Let* D_a *be a stable, stably embedded a-definable set. Then* D_a *can be defined by a formula* $\varphi(x, a)$ *with* $\varphi(x, y)$ *stable.*

PROOF. Let $p = \mathrm{tp}(a/\emptyset)$. Suppose that D_a is defined by the formula $\psi(x, a)$.

CLAIM. There do not exist elements a_i, b_i (for $i \in \omega$) such that $p(a_i)$ holds for each i and $b_i \in D_{a_j}$ if and only if $i > j$.

PROOF OF CLAIM. Suppose such a_i, b_i exist. Then $b_i \in D_{a_0}$ for all $i > 0$. Thus, $\psi(b_i, a_0) \wedge \psi(b_i, a_j)$ holds if and only if $i > j$ (for $i, j > 0$). It follows that the formula $\psi(x, a_0) \wedge \psi(x, y)$ is unstable. This contradicts the fact that D_{a_0} is stable and stably embedded.

Given the claim, it follows by compactness that there is a formula $\rho(y) \in p$ such that there do not exist a_i, b_i (for $i \in \omega$) such that $\rho(a_j)$ holds for all j and $\psi(b_i, a_j)$ if and only if $i > j$. Thus, the formula $\varphi(x, y) = \psi(x, y) \wedge \rho(y)$ is stable, and $\varphi(x, a)$ defines the same set as $\psi(x, a)$.   □

LEMMA 3.4. *Let* D_b *be a stable stably embedded Cb-definable set. Let* p *be a C-definable type, and assume that* $a \models p|Cb$ *and* $a \in D_b$. *Then there is a C-definable stable stably embedded set* D' *with* $a \in D'$.

PROOF. We may suppose that $C = \mathrm{acl}(C)$. For if such a set D' may be found over $\mathrm{acl}(C)$, then the union D'' of the C-conjugates of D' works over C.

By Lemma 3.3, there is a stable formula $\delta(x, y)$ such that D_b is defined by $\delta(x, b)$. Put $q := \mathrm{tp}(b/C)$. Let $\chi(x)$ be the formula $d_q y \delta(x, y)$, a formula over C. By forking symmetry for stable formulas (Remark 2.10), $\chi(a)$ holds. Also, by Remark 2.5, $\chi(x)$ is a (finite) positive Boolean combination of formulas $\delta(x, b_i)$ where $b_i \equiv_C b$ for each i. Each $\delta(x, b_i)$ defines a stable and stably embedded set, and hence so does $\chi(x)$. □

We shall say that St_C satisfies the condition $(*_C)$ if: whenever $b \in \mathrm{St}_C$, and D_b is a stable and stably embedded Cb-definable set, there is a formula $\psi(y) \in \mathrm{tp}(b/C)$ such that whenever $\psi(b')$ holds, $D_{b'}$ is stable and stably embedded.

LEMMA 3.5. *Suppose that* $(*_C)$ *holds, and* $C \subseteq C' \subset \mathrm{St}_C$. *Then every element of* $\mathrm{St}_{C'}$ *is interdefinable over* C *with an element of* St_C.

PROOF. Let $a \in \mathrm{St}_{C'}$. Then there is $b \in \mathrm{St}_C$ and a stable and stably embedded Cb-definable set D_b such that $a \in D_b$. Let $\psi(y)$ be the formula provided by $(*_C)$. As $b \in \mathrm{St}_C$, we may suppose that $\psi(y)$ defines a stable stably embedded set. Put $E := \{x : \exists y(\psi(y) \wedge x \in D_y)\}$, a C-definable set. Then $a \in E$, and by the type-counting criterion in Lemma 2.17, E is stable and stably embedded. □

REMARK 3.6. The condition $(*_C)$ holds for each C if St_B is $\mathrm{St}_\emptyset$-internal for each B. For suppose E is a sort of St_C, that $b \in E$, and that D_b is a stable stably embedded Cb-definable set. There is a stable stably embedded $\emptyset$-definable set D^* and some e-definable surjection $f_e : D^* \to D_b$. Now let $\psi(y)$ be the formula

$$y \in E \wedge \exists u(f_u \text{ is a surjection } D^* \to D_y).$$

If $\psi(y)$ then D_y is D^*-internal, so is stable and stably embedded.

As observed later in Remark 7.9 below, $(*_C)$ therefore holds in ACVF.

In ACVF, a family of k-vector spaces $\mathrm{VS}_{k,C}$ is defined over any base set C; see Chapter 7 below. (This modifies the notation of [12, Section 2.6], where we wrote $\mathrm{Int}_{k,C}$ for $\mathrm{VS}_{k,C}$). $\mathrm{VS}_{k,C}$ is a subset of St_C; while formally it is a proper subset, any C-definable set in St_C is in C-definable bijection with an element of $\mathrm{VS}_{k,C}$, so they can be viewed as the same.

We shall frequently in the text write statements like $\mathrm{tp}(A/C) \vdash \mathrm{tp}(A/CB)$. This means that for any A', if $A' \equiv_C A$ then $A' \equiv_{CB} A$. Observe that $\mathrm{tp}(A/C) \vdash \mathrm{tp}(A/CB)$ is equivalent to $\mathrm{tp}(B/C) \vdash \mathrm{tp}(B/CA)$. Indeed, suppose $\mathrm{tp}(A/C) \vdash \mathrm{tp}(A/CB)$, and let $B' \equiv_C B$. There is $g \in \mathrm{Aut}(\mathcal{U}/C)$ with $g(B') = B$. Then $g(A) \equiv_C A$, so $g(A) \equiv_{CB} A$. Thus there is $h \in \mathrm{Aut}(\mathcal{U}/CB)$ with $hg(A) = A$. Then $hg(B') = B$, so $B' \equiv_{CA} B$. This symmetry is used often without comment.

REMARK 3.7. The fact that St_C is stably embedded implies immediately that for any sets A, B, $\mathrm{tp}(B/CB^{\mathrm{st}}) \vdash \mathrm{tp}(B/CB^{\mathrm{st}}A^{\mathrm{st}})$ (equivalently, that $\mathrm{tp}(A^{\mathrm{st}}/CB^{\mathrm{st}}) \vdash \mathrm{tp}(A^{\mathrm{st}}/CB)$). To see this, let $B' \equiv_{CB^{\mathrm{st}}} B$. We must show $B' \equiv_{CB^{\mathrm{st}}A^{\mathrm{st}}} B$. Let $\varphi(x, a) \in \mathrm{tp}(B/CA^{\mathrm{st}}B^{\mathrm{st}})$ with $a \in A^{\mathrm{st}}$ and $\varphi(x, y)$ a formula over CB^{st}, and consider $\{y\colon \varphi(B, y) \text{ holds}\}$. This is a subset of St_C, defined with parameters from $C \cup B$. Since St_C is stably embedded, it must be definable by some formula $\psi(x, b')$, say, with parameters b' from St_C. The set of such parameters b' (an element of $\mathcal{U}^{\text{eq}} = \mathcal{U}$) is definable from B, hence is in $\text{dcl}(B) \cap \text{St}_C = B^{\text{st}} = B'^{\text{st}}$. It follows that $\{y\colon \varphi(B, y) \text{ holds}\}$ is CB^{st}-definable, so equals $\{y\colon \varphi(B', y) \text{ holds}\}$. Thus, $\varphi(B', a)$ holds, so $\varphi(x, a) \in \text{tp}(B'/CB^{\text{st}}A^{\text{st}})$, as required.

By the last paragraph we have in particular the following. Suppose g is an automorphism of $\mathcal{U}$ fixing CB^{st}. Then there is an automorphism h fixing $CA^{\text{st}}B^{\text{st}}$ such that $h(B) = g(B)$.

LEMMA 3.8. (i) $\text{tp}(B/CA^{\text{st}}) \vdash \text{tp}(B/CA)$ *if and only if* $\text{tp}(A/CB^{\text{st}}) \vdash \text{tp}(A/CB)$.

(ii) $\text{tp}(B/CA^{\text{st}}) \vdash \text{tp}(B/CA)$ *if and only if* $\text{tp}(B^{\text{st}}/CA^{\text{st}}) \cup \text{tp}(B/C) \vdash \text{tp}(B/CA)$.

PROOF. (i) Assume $\text{tp}(B/CA^{\text{st}}) \vdash \text{tp}(B/CA)$. Suppose g is an automorphism fixing CB^{st}. By Remark 3.7, there is an automorphism h fixing $CA^{\text{st}}B^{\text{st}}$ such that $h|_B = g|_B$. As $h^{-1}g$ fixes B, $g(A) \equiv_{CB} h(A)$. By our assumption and as h fixes CA^{st}, $B \equiv_{CA} h^{-1}(B)$, or equivalently, $A \equiv_{CB} h(A)$. Hence $A \equiv_{CB} g(A)$, as required. The other direction is by symmetry.

(ii) The right-to-left direction is immediate as $\text{tp}(B/CA^{\text{st}}) \vdash \text{tp}(B/C)$. For the left-to-right direction, suppose g is an automorphism fixing C and such that $A^{\text{st}}B^{\text{st}} \equiv_C A^{\text{st}}g(B^{\text{st}})$. Since also $A^{\text{st}}B^{\text{st}} \equiv_C g(A^{\text{st}})g(B^{\text{st}})$, there is an automorphism h fixing $g(B^{\text{st}})$ such that $h(g(A^{\text{st}})) = A^{\text{st}}$. By Remark 3.7, there is automorphism k fixing $A^{\text{st}}g(B^{\text{st}})$ such that $kh(g(B)) = g(B)$. Now khg fixes A^{st} and maps B to $g(B)$. Thus $B \equiv_{CA^{\text{st}}} g(B)$, so by the assumption, it follows that $B \equiv_{CA} g(B)$, as required. $\square$

DEFINITION 3.9. Define $A \underset{C}{\overset{d}{\smile}} B$ to hold if $A^{\text{st}} \underset{C}{\smile} B^{\text{st}}$ and $\text{tp}(B/CA^{\text{st}}) \vdash \text{tp}(B/CA)$. We say that $\text{tp}(A/C)$ is *stably dominated* if, whenever $B \subset \mathcal{U}$ and $A^{\text{st}} \underset{C}{\smile} B^{\text{st}}$, we have $A \underset{C}{\overset{d}{\smile}} B$.

REMARK 3.10. It follows from Proposition 3.32 (iii) below that if a is an infinite tuple, then $\text{tp}(a/C)$ is stably dominated if and only if $\text{tp}(a'/C)$ is stably dominated for every finite subtuple a' of a.

Symmetry of $\underset{}{\overset{d}{\smile}}$ follows immediately from Lemma 3.8 (i) and symmetry of stable non-forking applied in St_C.

LEMMA 3.11. $A \underset{C}{\overset{d}{\smile}} B$ *if and only if* $B \underset{C}{\overset{d}{\smile}} A$.

The definition of stable domination is analogous to that of compact domination introduced in [21], in the following sense: $\text{tp}(a/C)$ is stably dominated precisely if, for any definable set D, if $d^{\text{st}} := \text{St}_C(\ulcorner D \urcorner)$, then either $\{x \models p\colon x^{\text{st}} \underset{C}{\smile} d^{\text{st}}\} \subseteq D$, or $\{x \models p\colon x^{\text{st}} \underset{C}{\smile} d^{\text{st}}\} \cap D = \emptyset$. For compact domination, there is a similar definition, except that a^{st} is a finite tuple,

St$_C$ is a compact topological space, 'small' means 'of measure zero', and the uniformity condition on fibres is slightly weaker.

In many situations only a finite part of a^{st} is required for domination: that is, there is a C-definable map f to St$_C$ such that for any B, if $f(a) \underset{C}{\downarrow} B^{\mathrm{st}}$ then tp$(B/Cf(a)) \vdash$ tp(B/Ca). It will emerge that this is the situation in ACVF. In particular, the generic type of the valuation ring is stably dominated by that of the residue field: essentially, the type of a generic element of the valuation ring is dominated by its residue.

The following lemma will yield that stably dominated types have definable (so also invariant) extensions.

LEMMA 3.12. *Let M be a sufficiently saturated structure, and C a small subset of M. Let $p(x, y)$ be a type over M with x, y possibly infinite tuples, and let $q(x)$ be the restriction of p to the x-variables. Suppose that $p(x, y)$ is the unique extension over M of $p(x, y)|_C \cup q(x)$, and that q is C-definable. Then p is C-definable.*

PROOF. Suppose $\varphi(x, y, b) \in p$. Then there are formulas $\chi(x, b, c) \in q$ and $\psi(x, y, c) \in p|C$ such that $\chi(x, b, c) \wedge \psi(x, y, c) \vdash \varphi(x, y, b)$. Then the formula $\rho(x, b, c)$, namely $\chi(x, b, c) \wedge \forall y(\psi(x, y, c) \rightarrow \varphi(x, y, b))$, is in q. Thus, as q is C-definable, for the formula $d_q x \rho(x, y)$ over C we have $d_q x \rho(x, b)$. Thus, $\varphi(x, y, b) \in p$ if and only if for one of at most $|L(C)|$-many formulas $d_q x \rho$ over C, we have $d_q x \rho(x, b)$. Likewise, $\neg \varphi(x, y, b) \in p$ is equivalent to a disjunction of at most $|L(C)|$-many formulas over C about b. Definability of p now follows by compactness.   □

PROPOSITION 3.13. *Let M be a model, and $C \subset M$.*

(i) *Suppose that* tp(A/C) *is stably dominated. Then* tp(A/C) *is extendable to an* acl(C)-*definable (so* Aut$(M/\mathrm{acl}(C))$-*invariant) type over M, such that if A' realises this type then $A' \underset{C}{\overset{d}{\downarrow}} M$.*

(ii) *Suppose* tp(A/C) *is stably dominated and M is saturated, with $|A|, |C| < |M|$. Then* tp$(A/\mathrm{acl}(C))$ *has a unique* Aut$(M/\mathrm{acl}(C))$-*invariant extension.*

PROOF. (i) We may suppose that $A^{\mathrm{st}} \underset{C}{\downarrow} M^{\mathrm{st}}$. Then, by the hypothesis, tp$(A^{\mathrm{st}}A/M)$ is the unique extension of tp$(A^{\mathrm{st}}A/C) \cup$ tp(A^{st}/M) to M. Since tp$(A^{\mathrm{st}}/M^{\mathrm{st}})$ is definable over acl(C) in the stable structure St$_C$, it follows from Lemma 3.12 that tp(A/M) is definable over acl(C). In particular, tp(A/M) is Aut$(M/\mathrm{acl}(C))$-invariant.

(ii) Now suppose $A'' \equiv_{\mathrm{acl}(C)} A'$ and tp(A'/M), tp(A''/M) are both Aut$(M/\mathrm{acl}(C))$-invariant extensions of tp(A/C). Then tp$((A')^{\mathrm{st}}/M^{\mathrm{st}})$ and tp$((A'')^{\mathrm{st}}/M^{\mathrm{st}})$ are both invariant extensions of tp$(A^{\mathrm{st}}/\mathrm{acl}(C))$ in St$_C$: indeed, any automorphism in Aut$(\mathrm{St}_C/\mathrm{acl}(C))$ extends to an automorphism of M over acl(C) (saturation of M and stable embeddedness), so fixes tp(A'/M) and tp(A''/M), and hence fixes tp$((A')^{\mathrm{st}}/M^{\mathrm{st}})$ and tp$((A'')^{\mathrm{st}}/M^{\mathrm{st}})$. Hence tp$((A')^{\mathrm{st}}/M^{\mathrm{st}})$ and tp$((A'')^{\mathrm{st}}/M^{\mathrm{st}})$ are equal, as invariant extensions of a type over an algebraically closed base are unique in a saturated model of a stable theory (see, for example Remark 2.9). In particular, $(A')^{\mathrm{st}} \underset{C}{\downarrow} M^{\mathrm{st}}$. By stable domination, $\mathrm{tp}(A'/CM^{\mathrm{st}}) \vdash \mathrm{tp}(A'/CM)$. Thus, applying Lemma 3.8 (ii) with A', M replacing B, A, $\mathrm{tp}(A'/M) = \mathrm{tp}(A''/M)$.   □

REMARK 3.14. (i) In the proof of (i), $\mathrm{tp}(AA^{\mathrm{st}}/M^{\mathrm{st}})$ is definable over $C'' :=$ $\mathrm{acl}(C) \cap \mathrm{dcl}(CA)$. It follows that $\mathrm{tp}(A/C)$ extends to a C''-definable type over M. Likewise, in (ii), $\mathrm{tp}(A/C'')$ has a unique $\mathrm{Aut}(\mathcal{U}/C'')$-invariant extension.

(ii) By the proof of Proposition 3.13, if $C \subseteq B$ and $\mathrm{tp}(C/B)$ has an $\mathrm{Aut}(\mathcal{U}/\mathrm{acl}(C))$-invariant extension over $\mathcal{U}$, then $\mathrm{St}_C(A) \underset{C}{\overset{d}{\downarrow}} \mathrm{St}_C(B)$.

We will say that $\mathrm{Aut}(\mathcal{U}/C)$-invariant type p is *stably dominated* if $p|C$ is stably dominated.

In ACVF there are 'many' stably dominated types, enough in some sense to control the behaviour of all types, if Γ-indexed families are taken into account. We expect similar facts to hold in other theories of intended application, such as the rigid analytic expansions of Lipshitz [32] (see Ch. 16). Further examples are given in Chapter 16. One cannot however expect *all* types to be stably dominated, even over a model, unless the entire theory is stable.

COROLLARY 3.15. *Suppose that for each singleton a and every model M, $\mathrm{tp}(a/M)$ is stably dominated. Then the theory T is stable.*

PROOF. This follows immediately from Proposition 3.13 (i), since one of the standard characterisations of stability (Theorem 2.4 (iv)) is that every 1-type over a model is definable.   □

Next, we consider the relationship between indiscernible sequences and their trace in St_C. Recall that if $(a_i : i \in I)$ is an indiscernible sequence, and $J \subset I$, we write $a_J := \{a_j : j \in J\}$.

PROPOSITION 3.16. (i) *Let $(a_i : i \in I)$ be an infinite indiscernible sequence over C, with $\mathrm{acl}(Ca_J) \cap \mathrm{acl}(Ca_{J'}) = \mathrm{acl}(C)$ for any finite disjoint $J, J' \subset I$. Then $(\mathrm{St}_C(a_i) : i \in I)$ forms a Morley sequence in St_C.*

(ii) *Let q be an $\mathrm{Aut}(\mathcal{U}/C)$-invariant type, and suppose that the sequence $(a_i : i \in I)$ satisfies that for all $i \in I$, $a_i \models q|C \cup \{a_j : j < i\}$. Then $(a_i : i \in I)$ is C-indiscernible, and $\mathrm{acl}(Ca_J) \cap \mathrm{acl}(Ca_{J'}) = \mathrm{acl}(C)$ for any finite disjoint $J, J' \subset I$.*

PROOF. (i) Clearly, $(\mathrm{St}_C(a_i) : i \in I)$ is an indiscernible sequence (over C) in St_C, so by stability is an indiscernible set. That it is a Morley sequence now follows, for example, from Lemma 2.16.

(ii) The indiscernibility is straightforward. For the algebraic closure condition, suppose for a contradiction that there is $b \in (\mathrm{acl}(Ca_J) \cap \mathrm{acl}(Ca_{J'})) \setminus \mathrm{acl}(C)$, with $J \cap J' = \emptyset$. We may suppose that $J \cup J'$ is minimal subject to this (for b). Let $n := \max\{J \cup J'\}$, with $n \in J'$, say. Let M be a sufficiently saturated model containing $C \cup \{a_i : i < n\}$. We may suppose that $a_n \models q|M$. Then $b \in \mathrm{acl}(Ca_J) \subset M$, so $b \in M$. Also, the orbit of b under $\mathrm{Aut}(M/C \cup a_{J' \setminus \{n\}})$ is infinite, so there is $f \in \mathrm{Aut}(M/C \cup a_{J' \setminus \{n\}})$ such that $f(b) \notin \mathrm{acl}(Ca_{J'})$. Such f is not elementary over a_n. This contradicts the invariance of q.   □

PROPOSITION 3.17. *Let p be an $\mathrm{Aut}(\mathcal{U}/C)$-invariant type, with $p|C$ stably dominated. Then for any formula φ over $\mathcal{U}$, $\varphi \in p$ if and only if for some $\theta \in p|C$, $\theta \wedge \neg\varphi$ divides over C.*

PROOF. Write $\varphi = \varphi(x, b)$. If $\theta \wedge \neg\varphi$ divides over C, let $(b_j : j \in J)$ be an infinite sequence with $b = b_1$ and such that $\{\theta(x) \wedge \neg\varphi(x, b_j)\}$ is k-inconsistent. Let $a \models p|C(b_j : j \in J)$. Then $\theta(a)$ holds, so necessarily $\neg\varphi(a, b_j)$ fails for some j, i.e., $\varphi(a, b_j)$ holds. Thus $\varphi(x, b_j) \in p$. By $\mathrm{Aut}(\mathcal{U}/C)$-invariance, $\varphi \in p$.

Conversely, assume $\varphi \in p$. Let I be an index set, $|I| > |C| + |L|$, and let $(b_i : i \in I)$ be an indiscernible sequence over C, such that $b = b_1$ and $(\mathrm{St}_C(b_i) : i \in I)$ forms a Morley sequence over C in St_C. If $a \models p|C$, then by 2.11 $\mathrm{St}_C(a)$ is independent from some $\mathrm{St}_C(b_i)$ for some i; hence $a \models p|Cb_i$. Thus $p|C \cup \{\neg\varphi(x, b_i) : i \in I\}$ is inconsistent, and the proposition follows by compactness.   □

We will be concerned with issues of independence just in St_C itself, and of stationarity when moving from C to its algebraic closure. The rest of this chapter is primarily concerned with these stationarity issues. We therefore introduce the following notion of 'stationary' independence. We often use notation St_b rather than St_C, particularly when the base is varying.

DEFINITION 3.18. Suppose at least one of X, Y is a subset of St_b. We shall write $X \underset{b}{\overset{s}{\downarrow}} Y$ if $\mathrm{St}_b(X)$ and $\mathrm{St}_b(Y)$ are independent over b in St_b, and in addition,

(s1) for any $a \in \mathrm{acl}(b) \cap \mathrm{dcl}(Xb)$, $\mathrm{tp}(a/b) \vdash \mathrm{tp}(a/Yb)$.

If $\mathrm{dcl}(b) = \mathrm{acl}(b)$ then (s1) has no content. So in particular, $X \underset{\mathrm{acl}(b)}{\overset{s}{\downarrow}} Y$ means only that $\mathrm{St}_b(X)$, $\mathrm{St}_b(Y)$ are independent over $\mathrm{acl}(b)$ in the stable structure $\mathrm{St}_{\mathrm{acl}(b)}$, and hence also in St_b. (Recall that St_b has essentially the same domain as $\mathrm{St}_{\mathrm{acl}(b)}$.) The '$s$' in $\overset{s}{\downarrow}$ stands for 'stationary'.

PROPOSITION 3.19. *The following conditions are equivalent to (s1) (without any extra assumptions on b, X, Y, such as independence).*

(s2) *For any $a \in \mathrm{dcl}(Xb)$, $\mathrm{tp}(a/b) \vdash \mathrm{tp}(a/\mathrm{dcl}(Yb) \cap \mathrm{acl}(b))$.*

(s3) *For any $a \in \mathrm{acl}(b) \cap \mathrm{dcl}(Xb)$, $\mathrm{tp}(a/b) \vdash \mathrm{tp}(a/\mathrm{dcl}(Yb) \cap \mathrm{acl}(b))$.*

Assume now that $X \subset \mathrm{St}_b$ and X and $\mathrm{St}_b(Y)$ are independent in St_b. Then (s1)–(s3) are equivalent also to the following conditions.

(s4) *For any $a \in X$, $\mathrm{tp}(a/b)$ is stationary as far as $\mathrm{St}_b(Y)$; that is, if $a' \equiv_b a$ and a' and $\mathrm{St}_b(Y)$ are independent in St_b, then $a' \equiv_{\mathrm{St}_b(Y)} a$.*

(s5) *For any $a \in X$ and all finite sets of stable formulas Δ,* $\mathrm{tp}(a/b)$ *and* $\mathrm{tp}(a/\,\mathrm{St}_b(Y))$ *have the same Δ-rank and Δ-multiplicity.*

(s6) *For any $a \in X$, any stable formula φ, and any φ-type q consistent with* $\mathrm{tp}(a/\,\mathrm{St}_b(Y))$ *and definable over* $\mathrm{acl}(b)$, *the disjunction of all* $\mathrm{St}_b(Y)$-*conjugates of the q-definition of φ is defined over b.*

PROOF. (s1) $\Rightarrow$ (s2) Suppose $\sigma \in \mathrm{Aut}(\mathcal{U}/b)$. Let $a \in \mathrm{dcl}(Xb)$, and assume that $\{e_1, \ldots, e_r\}$ is some complete set of conjugates over b, with $e_1 \in \mathrm{dcl}(Yb)$. Suppose $\varphi(a, e_1)$ holds (φ over b). We must show $\varphi(\sigma(a), e_1)$ holds. So suppose without loss that

$$\{e_1, \ldots, e_s\} := \{x \in \{e_1, \ldots, e_r\}: \varphi(a, x)\}.$$

Let $e := \ulcorner\{e_1, \ldots, e_s\}\urcorner$. Then $e \in \mathrm{acl}(b) \cap \mathrm{dcl}(Xb)$. Hence $\sigma(e) \equiv_{Yb} e$. In particular e_1 is in the set coded by $\sigma(e)$, so $\varphi(\sigma(a), e_1)$.

(s2) $\Rightarrow$ (s3). This is immediate.

(s3) $\Rightarrow$ (s1). Let the conjugates of a over b be $a = a_1, \ldots, a_r$, and those over Yb be (without loss) $a_1, \ldots, a_s$. Suppose $a' \equiv_b a$. Suppose $\varphi(a, y)$ holds, where y is a tuple from Y and φ is over b. Since $\ulcorner\{a_1, \ldots, a_s\}\urcorner \in \mathrm{dcl}(Yb) \cap \mathrm{acl}(b)$, by (s3), a and a' have the same type over $b\ulcorner\{a_1, \ldots, a_s\}\urcorner$, so $a' \in \{a_1, \ldots, a_s\}$. Since $\varphi(a, y)$, also $\varphi(a_i, y)$ for $i = 1, \ldots, s$. Hence $\varphi(a', y)$ holds, as required.

Under the extra conditions, the equivalence of (s1) with (s4)–(s6) is an exercise in stability theory. $\square$

Note that (s2) is symmetric between X and Y. For assume (s2) and let $a' \in \mathrm{dcl}(Yb) \cap \mathrm{acl}(b)$. Then by (s2), $\mathrm{tp}(a'/b) \vdash \mathrm{tp}(a'/b\,\mathrm{dcl}(X))$, which is (s1) with X and Y reversed. Since (s1) implies (s2), (s2) also holds with X and Y reversed. It follows that the condition $X \underset{b}{\overset{s}{\downarrow}} Y$ is also symmetric.

We now have a sequence of easy lemmas giving basic properties of the stationarity condition. The first one is a slight extension of Proposition 3.13.

LEMMA 3.20. *Suppose a is a tuple in* St_b.

(i) $a \underset{\mathrm{acl}(b)}{\downarrow} \mathrm{St}_b(Y)$ (*in* St_b) *if and only if* $\mathrm{tp}(a/Yb)$ *extends to an* $\mathrm{Aut}(\mathcal{U}/\mathrm{acl}(b))$-*invariant type.*

(ii) $a \underset{b}{\overset{s}{\downarrow}} Y$ *if and only if* $\mathrm{tp}(a/Yb)$ *is the unique extension over Yb of* $\mathrm{tp}(a/b)$ *which extends to an* $\mathrm{Aut}(\mathcal{U}/\mathrm{acl}(b))$-*invariant type.*

(iii) *If p is an* $\mathrm{Aut}(\mathcal{U}/b)$-*invariant type and $a \models p|bY$, then $a \underset{b}{\overset{s}{\downarrow}} Y$.*

PROOF. (i) If $\mathrm{tp}(a/Yb)$ extends to an $\mathrm{Aut}(\mathcal{U}/\mathrm{acl}(b))$-invariant type p, then $p|\mathrm{St}_b$ is $\mathrm{Aut}(\mathrm{St}_b/\mathrm{acl}(b))$-invariant, and extends $\mathrm{tp}(a/\mathrm{St}_b(Y))$; hence $a, \mathrm{St}_b(Y)$ are independent in St_b (as noted in the proof of Proposition 3.13). Conversely, assume $a \underset{\mathrm{acl}(b)}{\downarrow} \mathrm{St}_b(Y)$. Then, as $a \subset \mathrm{St}_b$, $\mathrm{tp}(a/\mathrm{St}_b(Y))$ extends to an $\mathrm{Aut}(\mathrm{St}_b/\mathrm{acl}(b))$-invariant type p'. By stable embeddedness, since p' is a type in a sort of St_b, p' generates a complete type p over $\mathcal{U}$; so clearly p

is $\text{Aut}(\mathcal{U}/\text{acl}(b))$-invariant. Similarly, $\text{tp}(a/\text{St}_b(Y))$ implies $\text{tp}(a/bY)$, so p extends $\text{tp}(a/bY)$.

For (ii), it suffices as above to work within St_b, replacing Y by $\text{St}_b(Y)$. Both conditions imply $a \underset{b}{\downarrow} \text{St}_b(Y)$. The statement then is that $a \underset{b}{\overset{s}{\downarrow}} \text{St}_b(Y)$ if and only if $\text{tp}(a/b)$ is stationary as far as $\text{St}_b(Y)$. This is just the equivalence (s1) $\Leftrightarrow$ (s4) of Proposition 3.19.

(iii) Independence in St_b is as above. Apply Proposition 3.19(s3) for the stationarity condition. $\qquad\square$

The next lemma will be extended in Proposition 6.10.

LEMMA 3.21. *Let* $c \subset \text{St}_b$. *Then* $\text{St}_b(ac) = \text{dcl}(\text{St}_b(a)c) \cap \text{St}_b$.

PROOF. Let $E = \text{dcl}(\text{St}_b(a), c) \cap \text{St}_b$. Clearly $E \subseteq \text{St}_b(ac)$. We must show $\text{St}_b(ac) \subseteq E$. By Remark 3.7, $\text{tp}(a/\text{St}_b(a))$ implies $\text{tp}(a/\text{St}_b)$, so $\text{tp}(a/E)$ implies $\text{tp}(a/\text{St}_b)$; as $c \in E$, it follows that $\text{tp}(ac/E)$ implies $\text{tp}(ac/\text{St}_b)$ and in particular $\text{tp}(ac/E)$ implies $\text{tp}(ac/\text{St}_b(ac))$. Hence, $\text{tp}(\text{St}_b(ac)/E) \vdash \text{tp}(\text{St}_b(ac)/bac)$. As $\text{St}_b(ac) \subseteq \text{dcl}(bac)$, it follows that

$$\text{St}_b(ac) \subseteq \text{dcl}(E) \cap \text{St}_b = E. \qquad\square$$

The next technical proposition yields the natural properties of $\underset{}{\overset{s}{\downarrow}}$ given in Corollaries 3.23 and 3.24.

PROPOSITION 3.22. *Let* b *be a tuple from a set* Z, *and* $c \subset \text{St}_b$. *Then the following are equivalent.*

(i) $X \underset{Z}{\overset{s}{\downarrow}} c$.

(ii) $\text{St}_b(XZ) \underset{\text{St}_b(Z)}{\overset{s}{\downarrow}} c$.

(iii) $\text{St}_b(XZ) \underset{\text{St}_b(Z)}{\overset{s}{\downarrow}} c$ *holds in the many-sorted stable structure* St_b.

In case $Z \subset \text{St}_b$, (i)–(iii) *are also equivalent to each of*:

(iv) $\text{St}_b(X) \underset{Z}{\overset{s}{\downarrow}} c$

(v) $\text{St}_b(X) \underset{Z}{\overset{s}{\downarrow}} c$ *in the structure* St_b.

PROOF. Notice that $\text{St}_{\text{St}_b(Z)}(\text{St}_b(XZ)) \supseteq \text{St}_b(XZ)$.

(i) $\Rightarrow$ (iii). Assume (i). By Lemma 3.20 (i) and the symmetry of Definition 3.18, $\text{tp}(c/XZ)$ extends to an $\text{Aut}(\mathcal{U}/\text{acl}(Z))$-invariant type p. By stable embeddedness, for any W the restriction map $\text{Aut}(\mathcal{U}/W) \to \text{Aut}(\text{St}_b/\text{St}_b(W))$ is surjective. Hence, as $\text{St}_b(\text{acl}(Z)) \subseteq \text{acl}(\text{St}_b(Z))$, the type $p' := p|\text{St}_b$ is $\text{Aut}(\text{St}_b/\text{acl}(\text{St}_b(Z)))$-invariant. Also, p' extends $\text{tp}(c/\text{St}_b(XZ))$. Thus, in the structure St_b, we have $\text{St}_b(XZ) \underset{\text{acl}(\text{St}_b(Z))}{\downarrow} c$. To obtain (iii), we show that the symmetric form of condition (s1) of Definition 3.18 holds. If $d \in \text{acl}(\text{St}_b(Z)) \cap \text{dcl}(c\,\text{St}_b(Z))$, then $d \in \text{acl}(Z) \cap \text{dcl}(cZ)$, so by (i), $\text{tp}(d/Z)$ implies $\text{tp}(d/ZX)$. Now by stable embeddedness, $\text{tp}(d/\text{St}_b(Z))$

implies tp(d/Z). Hence, tp($d/St_b(Z)$) implies tp(d/ZX), which implies tp($d/St_{St_b(Z)}(St_b(XZ))$), as required.

(ii) $\Leftrightarrow$ (iii) The structure $St_{St_b(Z)}$ may be larger than St_b (it may have more sorts), but in the sorts of St_b, it has precisely the structure of St_b enriched with constants for $St_b(Z)$. So for the subsets $St_b(XZ)$ and c of St_b, independence and stationarity over $St_b(Z)$ have the same sense in the two structures.

(ii) $\Rightarrow$ (i). Assume (ii). Then by Lemma 3.20 (i), tp($c/St_b(XZ)$) extends to an Aut($\mathcal{U}/$acl($St_b(Z)$))-invariant type q; a fortiori, q is Aut($\mathcal{U}/$acl(Z))-invariant. As tp($c/St_b(XZ)$) implies tp(c/XZ) (stable embeddedness), q extends tp(c/XZ). Hence, by Lemma 3.20 (i), we have $X \underset{acl(Z)}{\overset{s}{\smile}} c$. To prove (i), we now use (twice) the second part of Lemma 3.20. Indeed, if q' is another Aut($\mathcal{U}/$acl(Z))-invariant extension of tp(c/Z), then (as in the proof of (i) $\Rightarrow$ (ii)) $q'|St_b$ is an Aut($St_b /$acl($St_b(Z)$))-invariant extension of tp($c/St_b(Z)$), so $q'|St_b(XZ) = q|St_b(XZ)$. Hence, since $c \in St_b$, $q'|XZ = q|XZ$ by stable embeddedness.

Finally, if $Z \subset St_b$, then $St_b(Z) = $dcl($Z$), and the equivalences (ii) $\Leftrightarrow$ (iv) and (iii) $\Leftrightarrow$ (v) are immediate, using Lemma 3.21. $\square$

COROLLARY 3.23 (Transitivity). *Assume that either* $ac \subset St_b$ *or* $d \subset St_b$. *Then* $ac \underset{b}{\overset{s}{\smile}} d$ *if and only if* $c \underset{b}{\overset{s}{\smile}} d$ *and* $a \underset{bc}{\overset{s}{\smile}} d$.

PROOF. First, observe that if $ac \in St_b$, then in all the relations, by definition, d can be replaced by $St_b(d)$. Thus, we may assume $d \subset St_b$. By Proposition 3.22 (i) $\Leftrightarrow$ (iii), we have to show, in St_b, that $St_b(abc) \underset{b}{\overset{s}{\smile}} d$ holds if and only if $St_b(bc) \underset{b}{\overset{s}{\smile}} d$ and $St_b(abc) \underset{St_b(bc)}{\overset{s}{\smile}} d$. This is just transitivity in the stable structure St_b (for the stationarity condition, it is easiest to use (s4)). $\square$

COROLLARY 3.24. *Assume* $ab \subset St_C$ *with* $a \underset{C}{\overset{s}{\smile}} b$ *and* $ab \underset{C}{\overset{s}{\smile}} e$. *Then* $a \underset{Ce}{\overset{s}{\smile}} b$.

PROOF. We use Corollary 3.23 repeatedly. From $ab \underset{C}{\overset{s}{\smile}} e$, this gives $a \underset{Cb}{\overset{s}{\smile}} e$, and hence $a \underset{Cb}{\overset{s}{\smile}} be$. By 3.23 again, since $a \underset{C}{\overset{s}{\smile}} b$, we have $a \underset{C}{\overset{s}{\smile}} be$, and hence $a \underset{Ce}{\overset{s}{\smile}} b$. $\square$

We introduce next a notion of dcl-*canonical base* appropriate for the relation $X \underset{Z}{\overset{s}{\smile}} Y$. See Section 2.3 above, and [40], for basics of canonical bases in stable theories.

DEFINITION 3.25. Let $Y \subset St_b$. Write Cb($X/Y; b$) for the smallest $Z = $dcl($Zb$) $\subset$ dcl(Yb) such that $X \underset{Z}{\overset{s}{\smile}} Y$ holds.

Note that by Proposition 3.22, as $Y \subset St_b$ and $Z = $dcl($Zb$) $\subset St_b$ then $X \underset{Z}{\overset{s}{\smile}} Y$ holds if and only if $St_b(X) \underset{Z}{\overset{s}{\smile}} Y$ in the stable structure St_b. Such dcl-canonical bases exist in stable structures; this is clearest perhaps from characterisation (s4) in Proposition 3.19, together with Section 4 (especially Example 4.3) of [39]. Thus, Cb($X/Y; b$) always exists (and is unique).

LEMMA 3.26. *Assume* $d \subset St_b$, $b' \underset{b}{\overset{s}{\downarrow}} d$, *and* $d' \in St_{b'}$ *where* d' *is a finite tuple. Assume* $d' \in \mathrm{dcl}(bb'd)$. *Then there exists a finite tuple* $f \in F :=$ $\mathrm{Cb}(b'd'/bd; b)$ *such that* $F = \mathrm{dcl}(bf)$.

PROOF. We have $d' \in \mathrm{dcl}(bb'F)$: indeed, $b'd' \underset{F}{\overset{s}{\downarrow}} d$, so $d' \underset{Fb'}{\overset{s}{\downarrow}} d$ (Corollary 3.23), and $d' \in \mathrm{dcl}(Fdb')$ (as $d' \in \mathrm{dcl}(bb'd)$), so $d' \in \mathrm{dcl}(Fb')$. Choose f to be any finite tuple in F such that $d' \in \mathrm{dcl}(bb'f)$. Then as $b' \underset{b}{\overset{s}{\downarrow}} d$ and $f \in \mathrm{dcl}(bd)$, we have $b' \underset{bf}{\overset{s}{\downarrow}} d$ (3.23 again), so $b'd' \underset{bf}{\overset{s}{\downarrow}} d$. By minimality of F, it follows that $F = \mathrm{dcl}(bf)$. $\qquad\square$

LEMMA 3.27. *Let* $b, d \in St_\emptyset$, *and let* a *be arbitrary. Assume* $a \underset{d}{\overset{s}{\downarrow}} b$ *and* $a \underset{b}{\overset{s}{\downarrow}} d$. *Let* $f := \mathrm{Cb}(a/b; \emptyset)$. *Then* $f \subseteq \mathrm{dcl}(d)$.

PROOF. Let $f' := \mathrm{Cb}(a/bd; \emptyset)$. By the minimality in the definition of canonical base, $f' \subseteq \mathrm{dcl}(d)$ (as $a \underset{d}{\overset{s}{\downarrow}} bd$). Similarly, $f' \subseteq \mathrm{dcl}(b)$. But we have $a \underset{f'}{\overset{s}{\downarrow}} bd$, so $a \underset{f'}{\overset{s}{\downarrow}} b$. Hence by minimality of $\mathrm{dcl}(f)$ it follows that $f \subseteq \mathrm{dcl}(f')$. Thus, $f \subseteq \mathrm{dcl}(d)$. $\qquad\square$

DEFINITION 3.28. By a *$*$-type* we mean a type in variables $(x_i)_{i\in I}$, where I is a possibly infinite index set. A *$*$-function* is a tuple $(f_i)_{i\in I}$ of definable functions. Definable functions always depend on finitely many variables, but may have other dummy variables. A $*$-function is *C-definable* if all its component functions are C-definable. We emphasise that in this chapter types have generally been infinitary. The purpose of this definition is rather to express how $*$-functions are decomposed into definable functions.

We say that a $*$-type p over C is *stably dominated via a C-definable $*$-function* f if, whenever $a \models p|C$, we have $f(a) \subset St_C$, and for any e, if $f(a) \underset{\mathrm{acl}(C)}{\overset{s}{\downarrow}} e$ (equivalently $f(a) \underset{\mathrm{acl}(C)}{\overset{s}{\downarrow}} St_C(e)$) then $\mathrm{tp}(e/Cf(a)) \vdash \mathrm{tp}(e/Ca)$.

It is clear that a $*$-type over C is stably dominated if and only if it is stably dominated by some C-definable $*$-function.

PROPOSITION 3.29. *Let* c *enumerate* $\mathrm{acl}(C)$. *Let* f *be a C-definable $*$-function from* $\mathrm{tp}(a/C)$ *to* St_C. *Then* $\mathrm{tp}(a/C)$ *is stably dominated via* f *if and only if both of the following hold*:
  (i) $\mathrm{tp}(a/Cf(a))$ *implies* $\mathrm{tp}(a/cf(a))$,
  (ii) $\mathrm{tp}(a/\mathrm{acl}(C))$ *is stably dominated via* f.

PROOF. Assume that $\mathrm{tp}(a/C)$ is stably dominated via f. So for any e, if $St_C(f(a)) \underset{c}{\overset{}{\downarrow}} St_C(e)$ then $\mathrm{tp}(e/Cf(a)) \vdash \mathrm{tp}(e/Ca)$. Putting $e = c$, we get $\mathrm{tp}(c/Cf(a)) \vdash \mathrm{tp}(c/Ca)$, which is equivalent to (i).

To see (ii), suppose $St_c(f(a)) \underset{c}{\overset{}{\downarrow}} St_c(e)$. Then $St_C(f(a)) \underset{c}{\overset{}{\downarrow}} St_C(e)$, so $St_C(f(a)) \underset{c}{\overset{}{\downarrow}} St_C(ec)$. Thus, $\mathrm{tp}(ec/Cf(a)) \vdash \mathrm{tp}(ec/Ca)$. In particular, $\mathrm{tp}(e/cf(a)) \vdash \mathrm{tp}(e/ca)$, which gives (ii).

Conversely, assume that (i) and (ii) hold. Suppose St$_C(f(a)) \underset{c}{\downarrow}$ St$_C(e)$. Then by (ii), tp$(e/cf(a))$ implies tp(e/ca). Thus tp$(a/cf(a))$ implies tp$(a/cf(a)e)$. But by (i), tp$(a/Cf(a))$ implies tp$(a/cf(a))$, so it implies tp$(a/cf(a)e)$ and in particular tp$(a/Cf(a)e)$, as required.   □

REMARK 3.30. The above proof also shows that if f is a C-definable ∗-function from tp(a/C) to St$_C$ and tp(a/C) is stably dominated by f, then for any C'' with $C \subseteq C'' \subseteq$ acl(C), tp(a/C'') is stably dominated by f.

COROLLARY 3.31. (i) *Let f be a C-definable ∗-function, and suppose that* tp$(a/$acl$(C))$ *is stably dominated via f. Then* tp(a/C) *is stably dominated via f', where $f'(a)$ enumerates $f(a) \cup ($acl$(C) \cap$ dcl$(Ca))$.*

(ii) *Suppose that* tp$(a/$acl$(C))$ *is stably dominated via F, where F is an* acl(C)-*definable ∗-function. Then* tp$(a/$acl$(C))$ *is stably dominated via a C-definable ∗-function.*

(iii) tp(a/C) *is stably dominated if and only if* tp$(a/$acl$(C))$ *is stably dominated.*

PROOF. Recall that the domains of St$_C$ and St$_{\text{acl}(C)}$ are the same, though there are more relations on St$_{\text{acl}(C)}$.

(i) In Proposition 3.29, condition (i) is equivalent to the condition acl$(C)\cap$ dcl$(Ca) =$ acl$(C)\cap$dcl$(Cf(a))$. This certainly holds if $f(a)$ contains acl$(C)\cap$ dcl(Ca). Thus, (i) follows from Proposition 3.29.

(ii) We may write $F(x) = F^*(d,x)$, where F^* is C-definable, and $d \in$ acl(C). Now define $f(x)$ to be a code for the ∗-function $d \mapsto F^*(d,x)$ (more precisely, a sequence of codes for the component functions): so $f(x)$ is a code for a ∗-function on the finite C-definable set of conjugates of d over C. Then tp$(a/$acl$(C))$ is stably dominated by f, which is C-definable.

(iii) By (i) and (ii), stable domination over acl(C) implies stable domination over C. The other direction is immediate from 3.29.   □

PROPOSITION 3.32. (i) *Suppose* tp(a/C) *is stably dominated, but* St$_C(a)=C$. *Then $a \subset$ dcl(C).*

(ii) *Suppose $B \subset$ St$_C$ and A is dominated by B over C (that is, whenever $B \underset{c}{\downarrow} D^{\text{st}}$ we have* tp$(D/CB) \vdash$ tp(D/CA)). *Then* tp(A/C) *is stably dominated.*

(iii) *Suppose* tp(A/C) *is stably dominated, and $a' \subset$ dcl(CA). Then* tp(a'/C) *is stably dominated.*

PROOF. (i) Suppose that $a \not\subset$ dcl(C) and choose a distinct conjugate a' of a over C. Then St$_C(a') = C$, so St$_C(a) \underset{c}{\downarrow}$ St$_C(a')$. Also, $a \equiv_{\text{St}_C(a)} a'$, so by stable domination $a \equiv_{Ca} a'$, which is a contradiction.

(ii) We may suppose $A^{\text{st}} \underset{C}{\downarrow} D^{\text{st}}$, and must show that tp$(D/CA^{\text{st}}) \vdash$ tp(D/CA). So suppose $D \equiv_{CA^{\text{st}}} D'$. It is possible to find $D'' \equiv_{\text{acl}(C)} D'$ with $D'' \equiv_{CA} D$. (Indeed, suppose $\{e_1,\dots,e_r\}$ is an orbit over C, and tp(D/CA) implies that $\varphi(D,e_i)$ holds if and only if $i = 1$; then $e_1 \in CA^{\text{st}}$, so

$\varphi(D', e_1)$ holds; thus the two conditions on D'' are consistent.) Observe that we may replace B by any $B' \equiv_{CA} B$, without affecting the assumption. By stable embeddedness, if $B' \equiv_{CA^{\mathrm{st}}} B$ then $B' \equiv_{CA} B$. Thus, we may choose B so that $B \underset{CA^{\mathrm{st}}}{\downarrow} (D''D')^{\mathrm{st}}$ in St_C. Since $A^{\mathrm{st}} \underset{C}{\downarrow} D''^{\mathrm{st}}$ and $A^{\mathrm{st}} \underset{C}{\downarrow} D'^{\mathrm{st}}$, by transitivity we have $B \underset{C}{\downarrow} D''^{\mathrm{st}}$ and $B \underset{C}{\downarrow} D'^{\mathrm{st}}$. Thus $D'' \equiv_{CB} D'$, so $D'' \equiv_{CA} D'$ by the domination assumption. It follows that $D \equiv_{CA} D'$.

(iii) We may apply (ii), since $\mathrm{tp}(a'/C)$ is dominated by A^{st}. $\square$

REMARK 3.33. It would be possible to localise the notion of stable domination by working with just some of the sorts of St_C. Let $\mathcal{F}_C$ be a collection of C-definable stable stably embedded sets, and let $\mathrm{St}_C^{\mathcal{F}}$ be the many-sorted structure whose sorts are the members of $\mathcal{F}$, with the induced C-definable structure. There will be a corresponding notion of $\mathcal{F}$-stable domination, defined as above but with $\mathrm{St}_C^{\mathcal{F}}$ replacing St_C. To develop a satisfactory theory, one would need certain closure properties of $\mathcal{F}$. We do not pursue this here.

CHAPTER 4

INVARIANT TYPES AND CHANGE OF BASE

The property of a type being stably dominated is quite sensitive to the base set of parameters. In this chapter, we will consider conditions under which stable domination is preserved under increasing or decreasing the base. Notice that it is not in general true that if $\mathrm{tp}(a/C)$ is stably dominated and $C \subset B$ then $\mathrm{tp}(a/B)$ is stably dominated. For example, in ACVF, in the notation of Part II (or [12]), let $C = \emptyset$, and let a be a generic element of the valuation ring. Let s be a code for the open ball $B_{<1}(a)$, and identify s with $\mathrm{res}(a)$. Then $\mathrm{tp}(a/C)$ is stably dominated by the residue map, but $\mathrm{tp}(a/Cs)$ is not stably dominated; in particular, $\mathrm{tp}(a/Cs)$ is not stably dominated by the residue map, since $s = \mathrm{res}(a)$ is already in the base Cs, so independence in St_{Cs} from s has no content. (This will become clearer in Part II; essentially, by Lemma 7.25 and Corollary 10.8, the generic type of a closed ball is stably dominated, but the generic type of an open ball is not.)

More generally, suppose every extension of $\mathrm{tp}(a/C)$ is stably dominated. Let $b = \mathrm{St}_C(a)$. Then $\mathrm{tp}(a/b)$ is stably dominated. But $\mathrm{St}_b(a) = b$, so $\mathrm{tp}(a/b)$ is stably dominated by b. Thus, by Proposition 3.32 (i), $a \in \mathrm{dcl}(b)$. Thus, under this assumption, a must be a tuple from St_C.

Thus, some further hypothesis is needed, and the one we use is that $\mathrm{tp}(a/B)$ has an $\mathrm{Aut}(\mathcal{U}/C)$-invariant extension to $\mathcal{U}$. This gives quite an easy result (Proposition 4.1 below) for the case of increasing the parameter set. To show that stable domination is preserved under decreasing the parameter set is rather more difficult. We give an easy special case in Lemma 4.2, and our most general result in Theorem 4.9. The hypothesis in 4.9 that $\mathrm{tp}(B/C)$ is invariant is stronger than it might be; it would be good to investigate the weakest assumptions under which some version of Theorem 4.9 could be proved.

First, we give the easy direction of base change.

PROPOSITION 4.1. *Suppose that* $C \subseteq B \subset \mathcal{U}$ *and that* p *is an* $\mathrm{Aut}(\mathcal{U}/\mathrm{acl}(C))$- *invariant type over* $\mathcal{U}$. *Suppose also that* $p|C$ *is stably dominated. Then* $p|B$ *is stably dominated.*

PROOF. Let $A \models p|B$. Suppose that $\mathrm{St}_B(A) \downarrow_B \mathrm{St}_B(D)$ (in St_B) and that $D \equiv_{\mathrm{St}_B(A)} D'$. We must show that $D \equiv_{BA} D'$.

As $A \models p|B$, $\mathrm{St}_C(A) \downarrow_C \mathrm{St}_C(B)$ (Lemma 3.20). Also, $\mathrm{St}_C(A) \downarrow_B$ $\mathrm{St}_C(BD)$ in St_B, so $\mathrm{St}_C(A) \downarrow_{\mathrm{St}_C(B)} \mathrm{St}_C(BD)$ in the structure St_C. Hence, by transitivity, $\mathrm{St}_C(A) \downarrow_C \mathrm{St}_C(BD)$ in St_C. The assumption $D \equiv_{\mathrm{St}_B(A)} D'$ gives $BD \equiv_{\mathrm{St}_C(A)} BD'$ (for $B \subset \mathrm{St}_B(A)$); so by stable domination of $p|C$, $BD \equiv_{CA} BD'$. Hence $D \equiv_{BA} D'$, as required.   □

We next tackle the harder (downwards) direction of base change, first dealing with an easy special case.

LEMMA 4.2. *Let g be a C-definable $*$-function, and let p, q be $\mathrm{Aut}(\mathcal{U}/C)$-invariant types. Assume that when $b \models q|C$, $p|Cb$ is stably dominated via g. Then $p|C$ is stably dominated via g.*

PROOF. First observe that by Lemma 3.4, if $a \models p|C$ then the entries a_i of $g(a)$ lie in St_C; for any such entry satisfies an $\mathrm{acl}(C)$-definable type q_i, and if $b \models q|C$ and $a_i \models p|Cb$, then a_i lies in an $\mathrm{acl}(C)b$-definable stable stably embedded set, and so lies in one definable over $\mathrm{acl}(C)$ and hence in one defined over C.

Since p is $\mathrm{Aut}(\mathcal{U}/C)$-invariant, $\mathrm{dcl}(Ca) \cap \mathrm{acl}(C) = \mathrm{dcl}(C)$. Thus, by Corollary 3.31 (i), it suffices to prove that if c is an enumeration of $\mathrm{acl}(C)$, then $p|c$ is stably dominated by g. Let $a \models p$ and suppose that for some e we have $g(a) \downarrow_c^s e$. We must show that $\mathrm{tp}(e/cg(a))$ implies $\mathrm{tp}(e/ca)$. So suppose that $e' \equiv_{cg(a)} e$. Let $b \models q|caee'$. Then $b \downarrow_{ce}^s g(a)$ (via Lemma 3.20 (iii) to obtain first $\mathrm{St}_{ce}(b) \downarrow_{ce}^s g(a)$). As $e \downarrow_c^s g(a)$, by transitivity, $be \downarrow_c^s g(a)$, and so $e \downarrow_{cb}^s g(a)$. Likewise, $e' \downarrow_{cb}^s g(a)$. By the choice of b, $e' \equiv_{cb} e$. It follows that $e' \equiv_{cbg(a)} e$ (use Proposition 3.19 (s4) and an automorphism, as $g(a) \in \mathrm{St}_C$). By domination over Cb and Remark 3.30, $\mathrm{tp}(e/cbg(a))$ implies $\mathrm{tp}(e/cba)$. Thus, $e' \equiv_{cba} e$, so $e' \equiv_{ca} e$.   □

In Theorem 4.9 we substantially extend this result, removing the assumption that stable domination over $p|Cb$ is via a fixed C-definable function.

LEMMA 4.3. *Let f_0, f_1, f_2 be C-definable $*$-functions, all with range in St_C, and let p be a type over C stably dominated via f_1 and via f_2. Suppose that for $a \models p$, $f_1(a) \downarrow_{Cf_0(a)}^s f_2(a)$. Also suppose $(*)$ that $\mathrm{acl}(C) \cap \mathrm{dcl}(Cf_0(a)) = \mathrm{acl}(C) \cap \mathrm{dcl}(Ca)$. Then p is stably dominated via f_0.*

PROOF. We prove the result over c, an enumeration of $\mathrm{acl}(C)$. For suppose the result is proved in this case. The assumption $(*)$ ensures that $\mathrm{tp}(a/Cf_0(a)) \vdash \mathrm{tp}(a/c)$. Hence, it will follow from Proposition 3.29 that p is stably dominated via f_0 over C.

Clearly, $f_1(a) \underset{cf_0(a)}{\overset{s}{\smile}} f_2(a)$, and by Proposition 3.29, any extension of p over c is stably dominated via f_1 and via f_2. Thus, we may now replace C by c.

Let $a \models p$, and let $a_i = f_i(a)$ for $i = 0, 1, 2$. So $a_1 \underset{ca_0}{\overset{s}{\smile}} a_2$.

CLAIM. For any d, if $d \underset{c}{\overset{s}{\smile}} a_0$ then $d \underset{c}{\overset{s}{\smile}} a_1$.

PROOF OF CLAIM. Since a_1, a_0 lie in St_c which is stably embedded, it suffices to prove the claim for $d \subset \mathrm{St}_c$. We now work with stable independence $\smile$ in the structure St_c. We may assume $d \underset{ca_0a_1}{\smile} a_2$. As $a_1 \underset{ca_0}{\smile} a_2$, by transitivity and symmetry we have $da_1 \underset{ca_0}{\smile} a_2$, so $d \underset{ca_0}{\smile} a_2$. But $d \underset{c}{\smile} a_0$, so $d \underset{c}{\smile} a_2$, and hence, as c is algebraically closed, $d \underset{c}{\overset{s}{\smile}} a_2$. By the domination assumption, $\mathrm{tp}(d/ca_2)$ implies $\mathrm{tp}(d/ca)$. In particular $\mathrm{tp}(d/ca_2)$ implies $\mathrm{tp}(d/ca_2a_1)$. Choose $d' \equiv_{ca_2} d$ with $d' \underset{ca_2}{\smile} a_1$. As $d \underset{c}{\smile} a_2$, also $d' \underset{c}{\smile} a_2$. By transitivity, $d' \underset{c}{\smile} a_1a_2$, so $d' \underset{c}{\overset{s}{\smile}} a_1a_2$. Hence also $d \underset{c}{\overset{s}{\smile}} a_1a_2$, and in particular $d \underset{c}{\overset{s}{\smile}} a_1$.

Now to prove the lemma, assume $e \underset{c}{\overset{s}{\smile}} a_0$. Then $e \underset{c}{\overset{s}{\smile}} a_1$ by the claim, so, as $\mathrm{tp}(a/c)$ is stably dominated by a_1, $\mathrm{tp}(e/ca_1) \vdash \mathrm{tp}(e/ca)$. Also, since $\mathrm{tp}(e/ca_0)$ implies $e \underset{c}{\overset{s}{\smile}} a_1$, it implies $\mathrm{tp}(e/ca_1)$, e.g., by Lemma 3.20 (ii). Thus $\mathrm{tp}(e/ca_0)$ implies $\mathrm{tp}(e/ca)$. $\square$

REMARK 4.4. If p extends to an $\mathrm{Aut}(\mathcal{U}/C)$-invariant type and $a \models p|\mathcal{U}$, then every elementary permutation of $\mathrm{acl}(C)$ which is the identity on C is elementary over a. In particular, $\mathrm{acl}(C) \cap \mathrm{dcl}(Ca) = \mathrm{dcl}(C)$. It follows that our assumption $(*)$ automatically holds in this case.

The next lemma gives a useful symmetry condition for invariant extensions. Without the stable domination assumption, any o-minimal structure $\mathcal{U}$ would provide a counterexample: let p and q both be the $\mathrm{Aut}(\mathcal{U})$-invariant type '$x < \mathcal{U}$'.

LEMMA 4.5. *Let p, q be $\mathrm{Aut}(\mathcal{U}/C)$-invariant types, with $p|C$ stably dominated. Let $a \models p|C, b \models q|Ca$. Then $a \models p|Cb$.*

PROOF. Since $p|C$ is stably dominated, we may suppose that $a \in \mathrm{St}_C$. By Lemma 3.20 (ii) it suffices to show $a \underset{C}{\overset{s}{\smile}} b$. By forking symmetry and Proposition 3.19 (s4), it suffices to show $\mathrm{St}_C(b) \underset{C}{\overset{s}{\smile}} a$. This follows from the assumption by Lemma 3.20 (ii) again. $\square$

LEMMA 4.6. *Let $((b_i, d_i): i \in I)$ be an indiscernible sequence, with $d_i \subset \mathrm{St}_{b_i}$ for each $i \in I$. Fix $i \in I$ and let J be an infinite subset of $\{i' \in I : i' < i\}$. Put $b_J := \{b_j : j \in J\}, d_J := \{d_j : j \in J\}$. Then $d_i \underset{b_Jb_id_J}{\overset{s}{\smile}} b_{<i}d_{<i}$.*

PROOF. We apply criterion (s5) in Proposition 3.19. We may suppose that d_i is a finite tuple. Let Δ be a finite set of formulas $\varphi(x, y, z)$, $y = (y_1, \ldots, y_n)$, such that $\varphi(x, y, b_j)$ implies $x \in \mathrm{St}_{b_j}$. Then $\varphi(x, y, b_j)$ is a stable formula. Let $\Delta_i = \{\varphi(x, y, b_i): \varphi(x, y, z) \in \Delta\}$. We have to show that (for any such Δ)

$$\mathrm{rk}_{\Delta_i} \mathrm{tp}(d_i/d_{<i}b_{\leq i}) = \mathrm{rk}_{\Delta_i} \mathrm{tp}(d_i/d_J b_J b_i)$$

and similarly for multiplicity. Now $\leq$ is clear. For the $\geq$-inequality, let $\psi \in \mathrm{tp}(d_i/d_{<i}b_{\leq i})$ have minimal Δ_i-rank and multiplicity. Then ψ uses finitely many constants from $d_{<i}$ and from $b_{<i}$ (as well as b_i). Since J is infinite, and by indiscernibility, these can be replaced by constants from d_J, b_J (keeping b_i) so that the resulting formula ψ' has the same Δ_i-rank. This gives the $\geq$ inequality and proves the lemma. $\square$

In the next lemma, we use the clumsy symbols i, J, J' for indices, instead of just 0,1,2. The reason is that in the proof of Lemma 4.8, this notation will fit with that of Lemma 4.6 (which will yield condition (A3) below). In applications, J, J' are sets of indices (much as in Lemma 4.6). The final two clauses of (A1) can be regarded as the first two with Ji in place of J (respectively $J'i$ in place of J'). And the final clause of (A2) is the first with $J \cup J'$ in place of J.

LEMMA 4.7. *Assume we are given* $b_J, b_{J'}, b_i, d_J, d_{J'}, d_i$, *with* $d_v \subset \mathrm{St}_{b_v}$ *for each* $v \in \{i, J, J'\}$, *satisfying the following independence conditions.*

$$d_J \underset{b_J}{\overset{s}{\downarrow}} b_{J'}, \quad d_{J'} \underset{b_{J'}}{\overset{s}{\downarrow}} b_J, \quad d_J d_i \underset{b_J b_i}{\overset{s}{\downarrow}} b_{J'}, \quad d_{J'} d_i \underset{b_{J'} b_i}{\overset{s}{\downarrow}} b_J \qquad (A1)$$

$$b_i \underset{b_J}{\overset{s}{\downarrow}} \mathrm{St}_{b_J}(b_J b_{J'} d_J d_{J'}), \quad b_i \underset{b_{J'}}{\overset{s}{\downarrow}} \mathrm{St}_{b_{J'}}(b_J b_{J'} d_J d_{J'}), \qquad (A2)$$

$$b_i \underset{b_J b_{J'}}{\overset{s}{\downarrow}} \mathrm{St}_{b_J b_{J'}}(b_J b_{J'} d_J d_{J'})$$

$$d_i \underset{b_J b_{J'} b_i d_J}{\overset{s}{\downarrow}} d_{J'}, \quad d_i \underset{b_J b_{J'} b_i d_{J'}}{\overset{s}{\downarrow}} d_J \qquad (A3)$$

Then the following hold.

(i)

$$b_i d_i \underset{b_J b_{J'} d_J}{\overset{s}{\downarrow}} d_{J'}.$$

(ii) *Let* f *enumerate* $\mathrm{Cb}(b_i d_i/b_J d_J; b_J)$, *and let* f' *enumerate* $\mathrm{Cb}(b_i d_i/b_{J'} d_{J'}; b_{J'})$. *Then* $\mathrm{dcl}(f b_J b_{J'}) = \mathrm{dcl}(f' b_J b_{J'})$.

PROOF. (i) By (A3), $b_i d_i \underset{b_J b_{J'} b_i d_J}{\overset{s}{\downarrow}} d_{J'}$. By forking symmetry,

$$d_{J'} \underset{b_J b_{J'} b_i d_J}{\overset{s}{\downarrow}} b_i d_i.$$

But by assumption (A2) (last clause),

$$d_{J'} \underset{b_J b_{J'} d_J}{\overset{s}{\downarrow}} b_i.$$

By transitivity, $d_{J'} \underset{b_J b_{J'} d_J}{\overset{s}{\downarrow}} b_i d_i.$

(ii) We have $b_i d_i \underset{b_J f}{\overset{s}{\smile}} d_J$, by definition of f. Thus by transitivity,

$$(1) \qquad\qquad d_i \underset{b_J b_i f}{\overset{s}{\smile}} d_J.$$

By (A1),

$$(1.1) \qquad\qquad d_i d_J \underset{b_J b_i}{\overset{s}{\smile}} b_{J'}.$$

We have $f \subset \mathrm{dcl}(b_J d_J)$, so $f d_i d_J \underset{b_J b_i}{\overset{s}{\smile}} b_{J'}$, and hence $d_i d_J \underset{b_J b_i f}{\overset{s}{\smile}} b_{J'}$. By Corollary 3.24,

$$(2) \qquad\qquad d_i \underset{b_{J'} b_J b_i f}{\overset{s}{\smile}} d_J.$$

By (A3),

$$(3') \qquad\qquad d_i \underset{b_J b_{J'} b_i d_J}{\overset{s}{\smile}} d_{J'}.$$

As $f \subset \mathrm{dcl}(b_J d_J)$, by (2), (3'), and transitivity,

$$(3) \qquad\qquad d_i \underset{b_J b_{J'} b_i f}{\overset{s}{\smile}} d_{J'}.$$

Now by (A2) (last clause)

$$b_i \underset{b_J b_{J'}}{\overset{s}{\smile}} d_J d_{J'},$$

so $b_i \underset{b_J b_{J'}}{\overset{s}{\smile}} d_{J'} f$, and thus $b_i \underset{b_{J'} b_J f}{\overset{s}{\smile}} d_{J'}$. From this, (3), and transitivity, we obtain

$$(4') \qquad\qquad b_i d_i \underset{b_J b_{J'} f}{\overset{s}{\smile}} d_{J'}.$$

Let $e = \mathrm{St}_{b_{J'}}(b_J f)$. Also, $b_J f \underset{b_{J'} e}{\overset{s}{\smile}} d_{J'}$ (for example, use Proposition 3.22 (iii) $\Rightarrow$ (i), with $X = b_J f, b = b_{J'}, Z = \mathrm{St}_{b'_J}(b_J f), c = d_{J'}$). Together with (4'), transitivity gives

$$(4) \qquad\qquad b_i d_i \underset{b_{J'} e}{\overset{s}{\smile}} d_{J'}.$$

Now by (A3), we have

$$(5') \qquad\qquad d_i \underset{b_J b_{J'} b_i d_{J'}}{\overset{s}{\smile}} d_J.$$

As $e \subset \mathrm{dcl}(b_{J'} b_J d_J)$,

$$(5) \qquad\qquad d_i \underset{b_J b_{J'} b_i d_{J'}}{\overset{s}{\downarrow}} e.$$

Dually to (1.1), we have $d_i d_{J'} \underset{b_{J'} b_i}{\overset{s}{\downarrow}} b_J$, so

$$(6) \qquad\qquad d_i \underset{b_{J'} b_i d_{J'}}{\overset{s}{\downarrow}} b_J.$$

By (5), (6) and transitivity, we obtain

$$(7) \qquad\qquad d_i \underset{b_{J'} b_i d_{J'}}{\overset{s}{\downarrow}} e.$$

By (A2), we have $b_i \underset{b_{J'}}{\overset{s}{\downarrow}} \mathrm{St}_{b_{J'}}(b_{J'} b_J d_J d_{J'})$. Thus, $b_i \underset{b_{J'}}{\overset{s}{\downarrow}} d_{J'} e$. Note that $e \subset \mathrm{St}_{b_{J'}}$, so Corollary 3.23 is applicable. Hence by transitivity $b_i \underset{b_{J'} d_{J'}}{\overset{s}{\downarrow}} e$, so by transitivity and (7),

$$(8) \qquad\qquad b_i d_i \underset{b_{J'} d_{J'}}{\overset{s}{\downarrow}} e.$$

Now f' enumerates the canonical base of $b_i d_i / b_{J'} d_{J'}$. It follows from (4), (8), and Lemma 3.27 (applied over $b_{J'}$) that $f' \subseteq \mathrm{dcl}(b_{J'} e) \subseteq \mathrm{dcl}(b_J b_{J'} f)$. Dually, $f \subseteq \mathrm{dcl}(b_J b_{J'} f')$. Thus, $\mathrm{dcl}(f b_J b_{J'}) = \mathrm{dcl}(f' b_J b_{J'})$. □

For the purposes of the lemma below, let p, q be $\mathrm{Aut}(\mathcal{U}/C)$-invariant ∗-types, and F a C-definable ∗-function. We will be interested in the property

$$\Pi^*(p, q, F)\colon \text{if } b \models q|C, b' \models q|Cb, a \models p|Cbb'$$

$$\text{then } \mathrm{dcl}(Cbb'F(a, b)) = \mathrm{dcl}(Cbb'F(a, b')).$$

If q is an $\mathrm{Aut}(\mathcal{U}/C)$-invariant ∗-type, let $q^\omega = \mathrm{tp}((b_0, b_1, \dots, b_\omega)/C)$ where $b_\alpha \models q|\mathrm{dcl}(Cb_\beta \colon \beta < \alpha)$. Then q^ω is also an $\mathrm{Aut}(\mathcal{U}/C)$-invariant ∗-type.

LEMMA 4.8. *Let p, q be $\mathrm{Aut}(\mathcal{U}/C)$-invariant ∗-types, $F = (F_\lambda)_{\lambda \in \Lambda}$ a ∗-definable function, with each F_λ C-definable. Assume that if $b \models q|C$, then $p|Cb$ is stably dominated via $F(x, b)$. Then there exists a C-definable ∗-function f such that:*
   (i) *if $b \models q^\omega |C$, then $p|Cb$ is stably dominated over Cb via $f(x, b)$;*
   (ii) $\Pi^*(p, q^\omega, f)$.
If already $\Pi^*(p, q, F)$, then we may choose f to have form $f = (f_\lambda)_{\lambda \in \Lambda}$, so that for each $\lambda \in \Lambda$, f_λ is a C-definable function and $\Pi^*(p, q^\omega, f_\lambda)$ holds.

PROOF. For convenience we now work over C, so drop all reference to C. Let $a \models p|\emptyset$. Let $I := \omega + \omega + 1$ and let $(b_i \colon i \in I)$ be an indiscernible sequence over a, such that $b_i \models q|ab_{<i}$ for each i (so obtained as in Proposition 3.16). Let $d_i(\lambda) := F_\lambda(a, b_i)$ for each i, λ, and put $d_i = (d_i(\lambda) \colon \lambda \in \Lambda)$. For each $\lambda \in \Lambda$, $(b_i d_i(\lambda) \colon i \in I)$ is an indiscernible sequence over a.

Let $J = \omega = (0, 1, 2, \dots)$, $J' = (\omega+n : n \in \omega) = (\omega, \omega+1, \omega+2, \dots)$, and $i = \omega 2 := \omega + \omega$. As in Lemma 4.6, let $b_J = \{b_j : j \in J\}$, and similarly $b_{J'}$, d_J, $d_{J'}$. Also, let $b^*_1 = (b_0, b_1, \dots, b_\omega)$, $b^*_2 = (b_{\omega+1}, b_{\omega+2}, \dots, b_{\omega 2})$, $d^*_1 = (d_0, d_1, \dots, d_\omega)$, $d^*_2 = (d_{\omega+1}, b_{\omega+2}, \dots, d_{\omega 2})$. The hypotheses of Lemma 4.6 hold. We shall verify that the hypotheses of Lemma 4.7 also hold.

A1 is easily seen via Lemma 4.5, which ensures $\mathrm{tp}(a/b_I) = p|b_I$. For instance $d_J \downarrow^s_{b_J} b_{J'}$ follows from $a \downarrow^s_{b_J} b_{J'}$. Condition A2 follows from the fact that $b_i \models q|b_{<i}a$, so $b_i \models q|\,\mathrm{St}_{b_J}(b_{<i}d_{<i})$, using Lemma 3.20 and invariance of q. The first clause of (A3) follows from Lemma 4.6, which yields $d_{\omega 2} \downarrow^s_{b_J b_{\omega 2} d_J} b_J b_{J'} d_J d_{J'}$, and hence $d_{\omega 2} \downarrow^s_{b_J b_{J'} b_{\omega 2} d_J} d_{J'}$. The second clause follows similarly.

Now $b^*_1 \models q^\omega$. This is an $\mathrm{Aut}(\mathcal{U})$-invariant type, and $b^*_2 \models q^\omega|b^*_1$.

Now $\mathrm{Cb}(b_{\omega 2} d_{\omega 2}/b_J d_J; b_J) \subseteq \mathrm{dcl}(ab_J)$. Thus, there is a $*$-function f such that $f(a, b^*_1)$ enumerates $\mathrm{Cb}(b_{\omega 2} d_{\omega 2}/b_J d_J; b_J)$. We may suppose f is a sequence of $*$-functions $(\hat{f}_\lambda)$, where $\hat{f}_\lambda(a, b^*_1)$ enumerates $\mathrm{Cb}(b_{\omega 2} d_{\omega 2}(\lambda)/b_J d_J; b_J)$. Here, the $d_{\omega 2}(\lambda)$ form a directed system of finite tuples with limit $d_{\omega 2}$, and each function $\hat{f}_\lambda$ depends on a and finitely many of the variables corresponding to $b_1, b_2, \dots$; notice that b_ω is not used. Likewise, by indiscernibility, $f(a, b^*_2)$ enumerates $\mathrm{Cb}(b_{\omega 2} d_{\omega 2}/b_{J'} d_{J'}; b_{J'})$. It follows by Lemma 4.7 (ii) that

$$\mathrm{dcl}(f(a, b^*_1), b_J, b_{J'}) = \mathrm{dcl}(f(a, b^*_2), b_J, b_{J'}).$$

This yields (ii) of the lemma.

Next, we obtain (i). By definition of canonical basis, $b_{\omega 2} d_{\omega 2} \downarrow^s_{f(a, b^*_1)} b_J d_J$, so $d_{\omega 2} \downarrow^s_{b_{\omega 2} b_J f(a, b^*_1)} d_J$. It follows from Lemma 4.3 and Remark 4.4, applied over $C' := \mathrm{dcl}(b_J b_{\omega 2})$, that $p|C'$ is stably dominated by $f(a, b^*_1)$. Hence, since $f(a, b^*_1)$ depends only on ab_J and $b_\omega, b_{\omega 2}$ have the same type over b_J, $p|b^*_1$ is stably dominated by $f(a, b^*_1)$, as required.

For the final assertion, we shall apply Lemma 3.26. We have $b_{\omega 2} \downarrow^s_{b_J} d_J$, as in the proof of (A1). Also, as $\Pi^*(p, q, F)$, we have $\mathrm{dcl}(b_1 b_{\omega 2} d_1) = \mathrm{dcl}(b_1 b_{\omega 2} d_{\omega 2})$. Hence $d_{\omega 2}(\lambda) \in \mathrm{dcl}(b_1 b_{\omega 2} d_1) \subset \mathrm{dcl}(b_J b_{\omega 2} d_J)$ for each λ. It follows from 3.26 that there is a finite tuple $f[\lambda] \in \hat{f}_\lambda(a, b^*_1) = \mathrm{Cb}(b_{\omega 2} d_{\omega 2}(\lambda)/b_J d_J; b_J)$ such that $\hat{f}_\lambda(a, b^*_1) = \mathrm{dcl}(b_J f[\lambda])$.

There is a $\emptyset$-definable function f_λ so that $f[\lambda] = f_\lambda(a, b_J)$. Thus, the function f defined earlier can be viewed as having components f_λ for $\lambda \in \Lambda$. We now apply Lemma 4.7 (ii) with $b_{\omega 2}$ as b_i and $f_\lambda(a, b_J) \in d_{\omega 2}$ as d_i. This yields that

$$\mathrm{dcl}(b_J b_{J'} f_\lambda(a, b_J)) = \mathrm{dcl}(b_J b_{J'} f_\lambda(a, b_{J'})).$$

Hence,

$$\mathrm{dcl}(b^*{}_1 b^*{}_2 f_\lambda(a, b_J)) = \mathrm{dcl}(b^*{}_1 b^*{}_2 f_\lambda(a, b_{J'})).\qquad\qquad\square$$

THEOREM 4.9. *Let p, q be* $\mathrm{Aut}(\mathcal{U}/C)$-*invariant ∗-types. Assume that whenever* $b \models q|C$, *the type $p|Cb$ is stably dominated. Then $p|C$ is stably dominated.*

PROOF. The strategy is to show that $p|Cb$ is stably dominated via a C-definable function g, and then to apply Lemma 4.2. As a first approximation of the proof, let $a \models p|Cb$, and put $d = f_b(a)$, where f_b is a Cb-definable function witnessing that $p|Cb$ is stably dominated. Define an equivalence relation E on $\mathrm{tp}(bd/C)$, putting $(b, d)E(b', d')$ if and only if, for $a \models p|Cbb'$, $f_b(a) = d \Leftrightarrow f_{b'}(a) = d'$. Then put $g(a) = \ulcorner (b, d)/E \urcorner$. The details are rather more intricate, and seem to require the previous lemma. The technical problems are: reducing to finite tuples which lie in St_C; identifying the right equivalence relation, and proof of transitivity; and proof that a code for the equivalence class lies in St_C.

Note at the outset that the symmetry Lemma 4.5 is true of p, q: if $b_0 \models q$, then $p|Cb_0$ is stably dominated, so 4.5 can be applied over b_0. Let $a \models p|Cb_0, b \models q|Cb_0 a$. Then $a \models p|Cb_0 b$. In particular, $a \models p|Cb$ and $b \models q|Ca$; that is, symmetry holds already over C. Likewise if b_λ is a subtuple of b and q_λ is the restriction of q to the corresponding variables, then $a \models p|Cb_\lambda$ and $b_\lambda \models q_\lambda|Ca$, so again symmetry holds.

The goal is to obtain a C-definable ∗-function g such that for $b \models q|C$, the type $p|Cb$ is stably dominated via g. We then apply Lemma 4.2.

We work over C, so omit all reference to C. There is a $\emptyset$-definable ∗-function F such that, if $b \models q|\emptyset$, then the type $p|b$ is stably dominated via $F(x, b)$. We apply Lemma 4.8 twice, first to obtain a function f' satisfying 4.8 (i) and (ii), and then, to use f' in place of F to obtain a new function f satisfying also the last assertion of 4.8. That is, we obtain an $\mathrm{Aut}(\mathcal{U}/\emptyset)$-invariant ∗-type q^ω (which we shall continue to write as q), and $f = (f_\lambda)_{\lambda \in \Lambda}$, f_λ a definable function, such that $p|b$ is stably dominated via $f(x, b)$ for any $b \models q|\emptyset$, and $\Pi^*(p, q, f_\lambda)$ for each $\lambda \in \Lambda$. It follows that for each λ there is a type q_λ (the restriction of the ∗-type q to a finite subset of the variables) such that f_λ depends only on the p-variables and q_λ-variables, and such that the following holds: if $b_\lambda \models q_\lambda|\emptyset$, $b'_\lambda \models q_\lambda|b_\lambda$, and $a \models p|b_\lambda b'_\lambda$, then

$$\mathrm{dcl}(b_\lambda b'_\lambda f_\lambda(a, b_\lambda)) = \mathrm{dcl}(b_\lambda b'_\lambda f_\lambda(a, b'_\lambda)).$$

This means that there is a $\emptyset$-definable function h_λ with inverse h_λ^{-1} such that if $b_\lambda \models q_\lambda|\emptyset$, $b'_\lambda \models q_\lambda|b_\lambda$, $a \models p|b_\lambda b'_\lambda$, then

$$(*)\quad h_\lambda(b_\lambda, b'_\lambda, f_\lambda(a, b_\lambda)) = f_\lambda(a, b'_\lambda) \text{ and } h_\lambda^{-1}(b_\lambda, b'_\lambda, f_\lambda(a, b'_\lambda)) = f_\lambda(a, b_\lambda).$$

Let $b \models q|\emptyset$, let b_λ be its restriction to the q_λ-variables, and suppose $a \models p|b$. Put $d_\lambda := f_\lambda(a, b_\lambda)$, and $d := (d_\lambda)_{\lambda \in \Lambda}$. Let Q be the set of realisations of $q|\emptyset$, Q_λ the set of realisations of $q_\lambda|\emptyset$, R be the set of realisations of $\mathrm{tp}(bd/\emptyset)$, and R_λ be the set of realisations of $\mathrm{tp}(b_\lambda d_\lambda/\emptyset)$.

If $b'_\lambda \models q_\lambda|b_\lambda$ and $a \models p|b_\lambda b'_\lambda$, then $f_\lambda(a, b'_\lambda) \underset{b'_\lambda}{\overset{s}{\smile}} b_\lambda$, since $\mathrm{tp}(f_\lambda(a, b'_\lambda)/b'_\lambda b_\lambda)$ extends to an $\mathrm{Aut}(\mathcal{U}/\emptyset)$-invariant type. Thus since R_λ is a complete type, we can add to ($*$) the following:

($*1$) if $(b_\lambda, d_\lambda) \in R_\lambda$, $(b_1)_\lambda \models q_\lambda|(b_\lambda, d_\lambda)$ then $h_\lambda(b_\lambda, (b_1)_\lambda, d_\lambda) \underset{(b_1)_\lambda}{\overset{s}{\smile}} b_\lambda$.

Since p is $\mathrm{Aut}(\mathcal{U}/b)$-invariant, $\mathrm{tp}(a/b)$ implies $\mathrm{tp}(a/\mathrm{acl}(b))$, so $\mathrm{tp}(d/b)$ implies $\mathrm{tp}(d/\mathrm{acl}(b))$. Thus, for any X, $d \underset{b}{\overset{s}{\smile}} X$ if and only if $d \underset{\mathrm{acl}(b)}{\overset{s}{\smile}} X$. The same holds with b, d replaced by b_λ, d_λ.

The domain of $h_\lambda(b_\lambda, b'_\lambda, x)$ contains $\{d_\lambda : b_\lambda d_\lambda \in R_\lambda \text{ and } d_\lambda \underset{b_\lambda}{\overset{s}{\smile}} b'_\lambda\}$. Also, as $\mathrm{tp}(a/b_\lambda b'_\lambda)$ has an $\mathrm{Aut}(\mathcal{U}/b'_\lambda)$-invariant extension, $h_\lambda(b_\lambda, b'_\lambda, d) \underset{b'_\lambda}{\overset{s}{\smile}} b_\lambda$, so we have:

($**$) the image of $h_\lambda(b_\lambda, b'_\lambda, x)$ (and the domain of $h_\lambda^{-1}(b_\lambda, b'_\lambda, y)$) contains

$$\{y : b'_\lambda y \in R_\lambda \text{ and } y \underset{b'_\lambda}{\overset{s}{\smile}} b_\lambda\}.$$

For each $\lambda \in \Lambda$, define an equivalence relation E_λ on R_λ: $(b_\lambda, d_\lambda)E_\lambda(b'_\lambda, d'_\lambda)$ if and only if, whenever $b''_\lambda \models q_\lambda|b_\lambda d_\lambda b'_\lambda d'_\lambda$, we have $h_\lambda(b_\lambda, b''_\lambda, d_\lambda) = h_\lambda(b'_\lambda, b''_\lambda, d'_\lambda)$. Until Claim 8 below, we are concerned only with $q_\lambda, b_\lambda, d_\lambda$, so for ease of notation we drop the subscript λ until then (except for $E_\lambda, R_\lambda, f_\lambda, h_\lambda$).

CLAIM 1. Assume $bd, b'd' \in R_\lambda, b_1 \models q_\lambda|bd$ and $b_1 \models q_\lambda|b'd'$. Then $(b, d)E_\lambda(b', d')$ if and only if $h_\lambda(b, b_1, d) = h_\lambda(b', b_1, d')$.

PROOF OF CLAIM. Let $d_1 = h_\lambda(b, b_1, d)$, $d'_1 = h_\lambda(b', b_1, d')$. Let $b_2 \models q_\lambda|bb'b_1dd'$. Let $d_2 = h_\lambda(b_1, b_2, d_1)$, $d'_2 = h_\lambda(b_1, b_2, d'_1)$. So $d_1 = h_\lambda^{-1}(b_1, b_2, d_2)$, $d'_1 = h_\lambda^{-1}(b_1, b_2, d'_2)$. Thus, $d_1 = d'_1$ if and only if $d_2 = d'_2$.

Thus, to prove the claim, we must show $(b, d)E_\lambda(b', d')$ if and only if $d_2 = d'_2$. To show this, by the definition of E_λ, is suffices to prove that $d_2 = h_\lambda(b, b_2, d)$ and $d'_2 = h_\lambda(b', b_2, d')$.

We shall prove this last assertion for d_2. Let $a^* \models p|bb_1b_2$, and let $d^* = f_\lambda(a, b)$. Then $d^* \in \mathrm{St}_b$, and $d^* \underset{b}{\overset{s}{\smile}} b_1 b_2$. Thus $\mathrm{tp}(d^*bb_1b_2/\emptyset) = \mathrm{tp}(dbb_1b_2/\emptyset)$. So there exists $a' \models p|bb_1b_2$ with $f_\lambda(a', b) = d$. By ($*$),

$$f_\lambda(a', b_1) = h_\lambda(b, b_1, f_\lambda(a', b)) = h_\lambda(b, b_1, d) = d_1,$$

and then

$$h_\lambda(b, b_2, d) = h_\lambda(b, b_2, f_\lambda(a', b)) = f_\lambda(a', b_2)$$
$$= h_\lambda(b_1, b_2, f_\lambda(a', b_1)) = h_\lambda(b_1, b_2, d_1) = d_2.$$

This and a similar argument for d'_2 yield the claim.

CLAIM 2. E_λ is a definable equivalence relation (i.e., the intersection with R_λ^2 of one).

PROOF OF CLAIM. By Claim 1, if $bd, b'd' \in R_\lambda$, then $(b, d)E_\lambda(b', d')$ if and only if there is b_1 such that

$$b_1 \models q_\lambda|bd, \; b_1 \models q_\lambda|b'd' \text{ and } h_\lambda(b, b_1, d) = h_\lambda(b', b_1, d').$$

Also, $\neg(b, d)E_\lambda(b', d')$ if and only if there is b_1 such that

$$b_1 \models q_\lambda|bd, b_1 \models q_\lambda|b'd' \text{ and } h_\lambda(b, b_1, d) \neq h_\lambda(b', b_1, d').$$

Thus both E_λ and $R_\lambda^2 \setminus E_\lambda$ are ∞-definable over $\emptyset$. The claim follows by compactness. (In more detail, by compactness, there is a definable relation E_λ' such that $E_\lambda' \cap R_\lambda^2 = E_\lambda$; and in general, if an equivalence relation on a complete type is induced by some definable relation, then it is induced by some definable *equivalence* relation on some definable set containing the complete type.)

CLAIM 3. Let $(b, d) \in R_\lambda$, $b' \in Q_\lambda$, and suppose $d \underset{b}{\overset{s}{\smile}} b'$. Then there exists d' with $(b, d)E_\lambda(b', d')$.

PROOF OF CLAIM. Let $b_1 \models q_\lambda|bb'd$, $d_1 = h_\lambda(b, b_1, d)$. By (*1) we have $d_1 \underset{b_1}{\overset{s}{\smile}} b$. Also $b_1 \underset{bb'}{\overset{s}{\smile}} d$ by genericity, so $b'b_1 \underset{b}{\overset{s}{\smile}} d$ by transitivity. Thus $b' \underset{bb_1}{\overset{s}{\smile}} d$, so $b' \underset{bb_1}{\overset{s}{\smile}} d_1$. But then $b' \underset{b_1}{\overset{s}{\smile}} d_1$ (by transitivity, since $d_1 \underset{b_1}{\overset{s}{\smile}} b$.) So by (**), d_1 is in the range of $h_\lambda(b', b_1, x)$. Hence there is $d' := h_\lambda^{-1}(b', b_1, h_\lambda(b, b_1, d))$. Now $b_1 \models q_\lambda|bd$ and $b_1 \models q_\lambda|b'd'$: to see the latter, note that since $b' \underset{b_1}{\overset{s}{\smile}} d_1$, if $a' \models p|b'b_1$ then $\text{tp}(f_\lambda(a', b_1)/b'b_1) = \text{tp}(d_1/b'b_1)$. Thus there exists $a \models p|b'b_1$ with $f_\lambda(a, b_1) = d_1$. Now $d' = h_\lambda^{-1}(b', b_1, f_\lambda(a, b_1)) = f_\lambda(a, b')$, and $b' \models q_\lambda|b_1a$, so $b_1 \models q_\lambda|b'd'$. Hence, by Claim 1, $(b, d)E_\lambda(b', d')$.

CLAIM 4. Let $(b, d), (b, d') \in R_\lambda$, with $(b, d)E_\lambda(b, d')$. Then $d = d'$.

PROOF OF CLAIM. Let $b' \models q_\lambda|bdd'$. Then by Claim 1, $h_\lambda(b, b', d) = h_\lambda(b, b', d') = d^*$, say. So $d = h_\lambda^{-1}(b, b', d^*) = d'$.

Now let $(b_i : i < |T|^+)$ be an indiscernible sequence, with $b_i \models q_\lambda|b_{<i}$ for each i.

CLAIM 5. Let $(b^*, d^*) \in R_\lambda$. Then there is $i < |T|^+$ and some d_i such that $(b_i, d_i) \in R_\lambda$ and $(b^*, d^*)E_\lambda(b_i, d_i)$.

PROOF OF CLAIM. Let $b \models q_\lambda|b_{<|T|^+}b^*d^*$. Then $b \underset{b^*}{\overset{s}{\smile}} d^*$. By Claim 3, $(b^*, d^*)E_\lambda(b, d)$ for some d. Now $\{\text{St}_b(b_i) : i < |T|^+\}$ forms a Morley sequence in St_b, by Proposition 3.16 (i) and (ii). So $b_i \underset{b}{\overset{s}{\smile}} d$ for some i, by properties of preweight (see Chapter 2). By Claim 3 again, there exists d_i with $(b_i, d_i)E_\lambda(b, d)$.

CLAIM 6. $R_\lambda/E_\lambda \subset \text{St}_\emptyset$.

PROOF OF CLAIM. By Claim 5, any element of R_λ/E_λ has the form $(b_i, d)/E_\lambda \in \mathrm{St}_{b_i}$, for some i. Hence, $R_\lambda/E_\lambda \subset \mathrm{St}_{\{b_i\,:\,i<|T|^+\}}$. Hence there is a $\emptyset$-definable subset of $\mathrm{St}_{\{b_i\,:\,i<|T|^+\}}$ which contains R_λ/E_λ. As it must be stable and stably embedded, the claim follows.

CLAIM 7. Let $b \models q_\lambda|\emptyset$. Let $a \models p|b$. Let $d = f_\lambda(a,b)$. Then $(b,d)/E_\lambda \in \mathrm{dcl}(a)$.

PROOF OF CLAIM. Let (b', d') be a-conjugate to (b,d). Then $d' = f_\lambda(a,b')$. Let $b'' \models q_\lambda|bb'a$. Then $h_\lambda(b, b'', d) = f(a, b'') = h_\lambda(b', b'', d')$, so $(b,d)E_\lambda(b',d')$.

Now let a, b_λ, d_λ be as in a, b, d of Claim 7 (so we revert to subscript notation), and put $g_\lambda(a) := \ulcorner (b_\lambda, d_\lambda)/E_\lambda \urcorner$. The following claim completes the proof of the proposition.

CLAIM 8. $p|\emptyset$ is stably dominated via $g = (g_\lambda \colon \lambda \in \Lambda)$.

PROOF OF CLAIM. As $a \models p|b$, symmetry yields $b \models q|a$. By Claim 4, $f_\lambda(a, b_\lambda)$ is the unique d_λ such that $g_\lambda(a) = \ulcorner (b_\lambda, d_\lambda)/E \urcorner$. Thus $f_\lambda(a, b_\lambda) \in \mathrm{dcl}(b_\lambda, g_\lambda(a))$. As $p|b$ is stably dominated via f, it is also stably dominated via g. By Lemma 4.2, $p|\emptyset$ is stably dominated. $\qquad\square$

COROLLARY 4.10. *Let T be a theory, $\mathcal{U}$ be a universal domain for T. Let Γ be a 0-definable stably embedded set with a 0-definable linear ordering, such that every type $\mathrm{tp}(\gamma/E)$ (with $\gamma \in \Gamma, E \subset \Gamma$) extends to an E - invariant type. Let f be a 0-definable function.*

(i) *Assume $\mathrm{tp}(a/B)$ extends to an $\mathrm{Aut}(\mathcal{U}/B)$-invariant type. Then so does $\mathrm{tp}(f(a)/B)$.*

(ii) *Assume $\mathrm{tp}(a/B, \Gamma(Ba))$ is stably dominated. Then so is $\mathrm{tp}(f(a)/B, \Gamma(B, f(a)))$.*

PROOF. (i) Let $\mathrm{tp}(a'/\mathcal{U})$ be an $\mathrm{Aut}(\mathcal{U}/B)$-invariant extension of $\mathrm{tp}(a/B)$. Then clearly $\mathrm{tp}(f(a')/\mathcal{U})$ is an $\mathrm{Aut}(\mathcal{U}/B)$-invariant extension of $\mathrm{tp}(f(a)/B)$.

(ii) Let γ enumerate $\Gamma(Ba)$, and γ' enumerate $\Gamma(B, f(a))$. Since $\mathrm{tp}(a/B, \gamma)$ is stably dominated, by Proposition 3.32 so is $\mathrm{tp}(f(a)/B, \gamma)$. But $\mathrm{tp}(f(a)/B, \gamma') \vdash \mathrm{tp}(f(a)/B, \Gamma)$ by stable embeddedness, and in particular $\mathrm{tp}(f(a)/B, \gamma') \vdash \mathrm{tp}(f(a)/B, \gamma)$. Let q be an $\mathrm{Aut}(\mathcal{U}/B, \gamma')$-invariant extension of $\mathrm{tp}(\gamma/B, \gamma')$; this exists by stable embeddedness of Γ. Then $\mathrm{tp}(\gamma/B(f(a))) = q|B(f(a))$, since both equal the unique extension of $\mathrm{tp}(\gamma/B(\gamma'))$ to $B(f(a))$. By Theorem 4.9, $\mathrm{tp}(f(a)/B, \gamma')$ is stably dominated. $\qquad\square$

Let T be a theory, with universal domain $\mathcal{U}$. Let S be a collection of sorts, and among them let Γ be a stably embedded sort with a 0-definable linear ordering.

DEFINITION 4.11. We say that S is *metastable* over Γ if for any finite product D of sorts of S, and any small $C \le \mathcal{U}$, we have:

(i) if $C = \mathrm{acl}(C)$, then for any $a \in D(\mathcal{U})$, $\mathrm{tp}(a/C)$ extends to an $\mathrm{Aut}(\mathcal{U}/C)$-invariant type;

(ii) for some small B with $C \subseteq B \leq \mathcal{U}$, for any $a \in D(\mathcal{U})$, $\mathrm{tp}(a/B, \Gamma(Ba))$ is stably dominated.

If S consists of all sorts, we say T is metastable over Γ.

In Theorem 12.18 we will show that algebraically closed valued fields are metastable.

COROLLARY 4.12. *Assume every sort of T lies in the definable closure of S, and S is metastable over Γ. Then T is metastable over Γ.*

PROOF. By assumption, any finite sequence of elements can be written as $f(a)$ for some 0-definable function f and some $a \in D$, where D is a product of sorts in S. The corollary is thus immediate from Corollary 4.10.   □

CHAPTER 5

A COMBINATORIAL LEMMA

We give in this chapter three versions of a combinatorial lemma which may be of independent interest. The proof of the first version uses Neumann's Lemma (stated below), and the proof of the second uses basic combinatorial facts about stability. The second version has content in any model and gives an explicit bound of $n/2$. Thirdly in Lemma 5.4 we lift the result from finite sets to stable sets. Our application of Lemma 5.1 will be in the next chapter, for the existence of 'strong' codes for germs of functions; $R(a, b)$ will be the set of points on which two functions with the same germ disagree.

LEMMA 5.1. *Let M be an ω-saturated structure with partial 1-types S, Q over $\emptyset$, and with a $\emptyset$-definable relation R_0 inducing a relation $R \subseteq Q^2 \times S$. Assume*
(i) *for any $a, b \in Q$, $R(a, b) := \{z \in S : R(a, b, z)\}$ is a finite subset of S, and $R(a, b) = R(b, a)$;*
(ii) *for all $a, b, c \in Q$, $R(a, c) \subseteq R(a, b) \cup R(b, c)$*
(iii) *for all $a \in Q$, $R(a, a) = \emptyset$.*
Then for all $a, b \in Q$ we have $R(a, b) \subseteq \text{acl}(a) \cup \text{acl}(b)$.

PROOF. By the saturation assumption, $R(a, b)$ is a finite subset of S of bounded size. Hence (i)-(iii) remain true in any elementary extension, while the conclusion clearly descends. So we may assume M is ω-homogeneous.

We may also assume $\text{acl}(\emptyset) \cap S = \emptyset$, and that $Q \not\subseteq \text{acl}(\emptyset)$. For the lemma is trivial if $Q \subseteq \text{acl}(\emptyset)$. If $\text{acl}(\emptyset) \cap S \neq \emptyset$, we may replace S by $S' := S \setminus \text{acl}(\emptyset)$ and R by $R' := R \cap (Q^2 \times S')$. Then (i), (ii) still hold, and the conclusion for R' implies that for R.

CLAIM 1. For any finite $E, C \subset M$ and any finite tuple $d \in M$, there is an $\text{Aut}(M/C)$-translate d' of d with $\text{acl}(d'C) \cap \text{acl}(EC) = \text{acl}(C)$.

PROOF OF CLAIM. This follows by compactness from Neumann's Lemma [38, Lemma 2.3], which states that if G is a permutation group on an infinite set X with no finite orbits, then any finite set $F \subset X$ has some (and hence infinitely many) disjoint translates.

Now let $\mathbf{Q}_2 := \{(a, b) \in Q^2 : \text{acl}(a) \cap \text{acl}(b) = \text{acl}(\emptyset)\}$.

CLAIM 2. It suffices to prove the lemma for $(a, b) \in \mathbf{Q}_2$.

PROOF OF CLAIM. Suppose the lemma holds for elements of $\mathbf{Q}_2$, and let $a, b \in Q$ be distinct. By Claim 1, there is $c \in Q$ with $\mathrm{acl}(c) \cap \mathrm{acl}(a, b) = \mathrm{acl}(\emptyset)$. Hence $(a, c), (b, c) \in \mathbf{Q}_2$, so $R(a, c) \subseteq \mathrm{acl}(a) \cup \mathrm{acl}(c)$ and $R(b, c) \subseteq \mathrm{acl}(b) \cup \mathrm{acl}(c)$. Thus,

$$R(a, b) \subseteq R(a, c) \cup R(b, c) \subseteq \mathrm{acl}(a) \cup \mathrm{acl}(b) \cup \mathrm{acl}(c).$$

Also, $R(a, b) \subseteq \mathrm{acl}(a, b)$, so $R(a, b) \cap \mathrm{acl}(c) = \mathrm{acl}(\emptyset)$. Thus, $R(a, b) \subseteq \mathrm{acl}(a) \cup \mathrm{acl}(b)$, as required.

Since the size of $R(a, b)$ is bounded, we can proceed by reverse induction. So suppose that $R(a, b) \subseteq \mathrm{acl}(a) \cup \mathrm{acl}(b)$ whenever $(a, b) \in \mathbf{Q}_2$ and $|R(a, b)| > n$. We must prove $R(a, b) \subseteq \mathrm{acl}(a) \cup \mathrm{acl}(b)$ when $(a, b) \in \mathbf{Q}_2$ and $|R(a, b)| = n$. Suppose for a contradiction that this is false for some $(a, b) \in \mathbf{Q}_2$.

Put $m := |\mathrm{acl}(b) \cap R(a, b)|$ and $m' := |\mathrm{acl}(a) \cap R(a, b)|$. We will show in Claim 3 that $m \geq n/2$. A symmetrical argument will give $m' \geq n/2$. Since $(a, b) \in \mathbf{Q}_2$ and $S \cap \mathrm{acl}(\emptyset) = \emptyset$, we have $\mathrm{acl}(a) \cap \mathrm{acl}(b) \cap S = \emptyset$. By counting it will follow from Claim 3 that $R(a, b) \subseteq \mathrm{acl}(a) \cup \mathrm{acl}(b)$, contradicting our assumption on a, b.

CLAIM 3. $m \geq n/2$.

PROOF. i[Proof of Claim] By Claim 1 applied over b, there is $a' \equiv_b a$ with $\mathrm{acl}(a'b) \cap \mathrm{acl}(ab) = \mathrm{acl}(b)$. Then

$$\mathrm{acl}(a) \cap \mathrm{acl}(a') \subseteq \mathrm{acl}(a) \cap \mathrm{acl}(ab) \cap \mathrm{acl}(a'b) \subseteq \mathrm{acl}(a) \cap \mathrm{acl}(b) = \mathrm{acl}(\emptyset),$$

so $(a, a') \in \mathbf{Q}_2$.

If $|R(a, a')| > n$, then by induction we have $R(a, a') \subseteq \mathrm{acl}(a) \cup \mathrm{acl}(a')$. Then

$$R(a, b) \subseteq R(a, a') \cup R(a', b) \subseteq \mathrm{acl}(a) \cup \mathrm{acl}(a') \cup R(a', b).$$

But $R(a, b) \subseteq \mathrm{acl}(ab)$. So

$$R(a, b) \subseteq (\mathrm{acl}(a) \cup \mathrm{acl}(a'b)) \cap \mathrm{acl}(ab) \subseteq \mathrm{acl}(a) \cup \mathrm{acl}(b),$$

again contradicting the assumption on a, b.

Thus, we may suppose $|R(a, a')| \leq n$. However, by (ii), $R(a, a')$ contains the symmetric difference of $R(a, b), R(a', b)$. We have $|R(a, b)| = |R(a', b)| = n$, since $(a, b), (a', b)$ are conjugate. Also,

$$R(a, b) \cap R(a', b) \subseteq R(a, b) \cap (\mathrm{acl}(ab) \cap \mathrm{acl}(a'b)) \subseteq R(a, b) \cap \mathrm{acl}(b),$$

so $|R(a, b) \cap R(a', b)| \leq m$. Thus, the symmetric difference of $R(a, b), R(a', b)$ has size at least $2(n - m)$. Hence $n \geq 2(n - m)$, so $m \geq n/2$, as required. $\square$

In the second version below, if $T \subseteq Q \times S$ and $a \in Q$ then $T(a) := \{y \in S : (a, y) \in T\}$. Likewise, if $a, b \in Q$ then $R(a, b) := \{y \in S : R(a, b, y)\}$.

LEMMA 5.2. *Let* $n \in \mathbb{N}$. *Then there is an integer* n_0 *depending on* n *with the following property. Let* Q, S *be disjoint sets, and* $R \subseteq Q^2 \times S$. *Suppose that for all* $a, b, c \in Q$, *we have* $R(a, b) = R(b, a) \subseteq R(a, c) \cup R(b, c)$, *and* $|R(a, b)| \leq n$ *and* $R(a, a) = \emptyset$. *Then there is* $S_0 \subseteq S$ *with* $|S_0| \leq n_0$ *and* $T \subseteq Q \times S$, *with* $|T(a)| \leq n/2$ *for all* $a \in Q$ *such that, letting* $S' = S \setminus S_0$, $R' = R \cap (Q^2 \times S')$,

$$T(a) \triangle T(b) \subseteq R'(a, b) \subseteq T(a) \cup T(b)$$

for all $a, b \in Q$. *Moreover,* S_0 *and* T *are defined in the structure* $(Q, S; R)$ *by parameter-free first-order formulas, depending only on* n.

PROOF. We shall prove the result in a fixed (possibly infinite) structure $M = (Q, S; R)$. Compactness then implies that the formulas and bounds are uniform.

CLAIM 1. Let $N = n + 2$. Assume there are $a_i, b_i, c_i \in M$ for $i = 0, \ldots, N$ such that for all $j = 0, \ldots, N$, $R(a_i, b_j, c_i)$ holds whenever $i < j$. Then $R(a_i, b_j, c_i)$ holds for some i, j with $i \geq j$.

PROOF OF CLAIM. Suppose otherwise. We first argue that $c_0, \ldots, c_N$ are distinct. To see this, suppose that $c_i = c_j$ where $i < j$. Then $c_j \in R(a_i, b_j)$, and $c_j \notin R(a_i, b_i)$. As $R(a_i, b_j) \subseteq R(a_i, b_i) \cup R(b_i, b_j)$, it follows that $c_j \in R(b_i, b_j)$. Now $R(b_i, b_j) \subseteq R(a_j, b_i) \cup R(a_j, b_j)$, so $c_j \in R(a_j, b_i)$ or $c_j \in R(a_j, b_j)$, and each of these contradicts our assumption.

Whenever $j' < i < j$ we have $c_i \in R(b'_j, b_j)$: indeed, $c_i \in R(a_i, b_j) \subseteq R(a_i, b_{j'}) \cup R(b_j, b_{j'})$ but $c_i \notin R(a_i, b_{j'})$. Thus $c_i \in R(b_0, b_N)$ for all i with $0 < i < N$. As the c_i are distinct, this contradicts the assumption that $|R(b_0, b_N)| \leq n$.

Let $x = (x_1, x_2)$ be a variable ranging over $Q \times S$, let y range over Q, and let $\varphi(x, y) = R(x_1, y, x_2)$. Then Claim 1 states that φ is a stable formula. Let $q_1(y), \ldots, q_m(y)$ be the φ-types of maximal $\{\varphi\}$-rank in the variable y. For $j = 1, \ldots, m$, put $T_j = \{(x_1, x_2) \in Q \times S : (d_{q_j} y)\varphi(x_1 x_2, y)\}$. Let $T' = T_1 \cup \cdots \cup T_m$. As φ is stable, the q_i are definable φ-types (see Remark 2.10), so the T_i and T' are definable. Observe that as $|T_j(a)| \leq n$ for all a, $|T'(x)|$ is bounded as x varies through Q.

CLAIM 2. If $a, b \in Q$ then $R(a, b) \subseteq T'(a) \cup T'(b)$.

PROOF. It suffices to show that for each j, if $c \in R(a, b)$ then $c \in T_j(a) \cup T_j(b)$. To see this, choose $d \models q_j | \{a, b, c\}$. Since $R(a, b) \subseteq R(a, d) \cup R(b, d)$, $c \in R(a, d) \cup R(b, d)$. We suppose $c \in R(a, d)$. Then $\varphi(ac, d)$ holds, so by the choice of d, $(a, c) \in T_j$, so $c \in T_j(a)$. Similarly, if $c \in R(b, d)$ then $c \in T_j(b)$.

CLAIM 3. If $a, b \in Q$ then the symmetric difference $T'(a) \triangle T'(b)$ is a subset of $R(a, b)$.

PROOF. Let $c \in T'(a) \triangle T'(b)$, say with $c \in T_j(a) \setminus T'(b)$. Let $d \models q_j | \{a, b, c\}$. Then $\varphi(ac, d)$ (as $c \in T_j(a)$) and $\neg\varphi(bc, d)$ (as $c \notin T_j(b)$). So $c \in R(a, d)$ and $c \notin R(b, d)$. But $R(a, d) \subseteq R(b, d) \cup R(a, b)$. So $c \in R(a, b)$.

We now choose S_0 to be a finite $\emptyset$-definable subset of S, chosen so as to minimise $\ell := \text{Max}\{|T'(x) \setminus S_0| : x \in Q\}$. Define $T := T' \cap (Q \times (S \setminus S_0))$, so that $T(a) = T'(a) \setminus S_0$ for each a. Then T is $\emptyset$-definable. By Claims 2 and 3,

$$T(a) \triangle T(b) \subseteq R'(a, b) \subseteq T(a) \cup T(b)$$

for all $a, b \in Q$. To complete the proof, we will show that $\ell \leq n/2$.
Let

$$H_0 = \{w \subseteq S : |w| \leq \ell, \text{ for any } a' \in Q \text{ with } |T(a')| = \ell, w \cap T(a') \neq \emptyset\}.$$

Let H be the set of minimal elements of H_0 (under inclusion). Then H is finite. For otherwise, there exists $H' \subseteq H$, $|H'| > \ell$, forming a Δ-system, that is, there is some $w_0 \subseteq S$ such that any two distinct elements of H' have intersection w_0. Any $T(a')$ with $|T(a')| = \ell$ must meet any element of H' nontrivially; but it can meet at most m of the disjoint sets $w \setminus w_0$; so it must meet w_0 nontrivially. But then $w_0 \in H_0$, contradicting the minimality in the definition of H. In particular, the $\emptyset$-definable set $\bigcup H$ is finite (possibly empty).

There is $a \in Q$ such that $|T(a)| = \ell$ and $T(a) \cap \bigcup H = \emptyset$; otherwise, $|T(a) \setminus \bigcup H| < \ell$ for all $a \in Q$, contradicting the minimality in the choice of ℓ. Hence $T(a) \notin H_0$, as otherwise $T(a)$ contains some member of H. So by definition of H_0 there exists $a' \in Q$ such that $|T(a')| = \ell$ and $T(a) \cap T(a') = \emptyset$. But then, by Claim 3, $R(a, a') \supseteq T(a) \cup T(a')$, so $R(a, a') \geq 2\ell$. It follows that $\ell \leq n/2$.   □

REMARK 5.3. 1. In Lemma 5.2, the formulas defining T and S_0 can also be taken to be quantifier-free (in the structure $(Q, S; R)$), but with parameters. For S_0 this is immediate, as S_0 is finite, and for T_0 it holds since each T_j is quantifier-free definable (by a 'majority rule' definition for the average of the appropriate indiscernible sequence, see [40].)
2. It is easy to deduce Lemma 5.1 from Lemma 5.2, by compactness.

LEMMA 5.4. *Let M be an ω-saturated structure, Δ a finite set of stable formulas. Let Q be a partial type over $\emptyset$, S' a $\emptyset$-definable set, and X_0 an $\emptyset$-definable relation inducing a relation $X \subseteq Q^2 \times S'$.*
For $a, b \in Q$, let $X(a, b) := \{z \in S' : X(a, b, z)\}$. Let $r \in \mathbb{N}$. Assume:

(i) *for any $a, b \in Q$, $\text{rk}_\Delta(X(a, b)) \leq r$.*
(ii) *for all $a, b, c \in Q$, $X(a, c) \subseteq X(a, b) \cup X(b, c)$.*
(iii) *for all $a, b \in Q$, $X(a, b) = X(b, a)$; and $X(a, a) = \emptyset$.*

Then there exists a $\emptyset$-definable set D_0 inducing $D \subseteq Q \times S'$ such that
(i') *for any* $a \in Q$, $\mathrm{rk}_\Delta(D(a)) \leq r$.
(ii') *for all* $a, b \in Q$ *we have* $X(a, b) \subseteq D(a) \cup D(b)$.

PROOF. We may assume M is ω-homogeneous. For $r = 0$, the statement amounts to Lemma 5.1. We proceed by induction on r. We may suppose there are $b, b' \in Q$ with $\mathrm{rk}_\Delta(X(b, b')) = r$.

Let S be the set of all Δ-types over M of rank r, in the sort S'; we may regard S as a set of canonical parameters of these types. For any $b, b' \in Q$, among the Δ-types consistent with $X(b, b')$, let $R(b, b')$ be the set of Δ-types over M consistent with $X(b, b')$, and of Δ-rank r; so $R \subset Q^2 \times S$, the setting of Lemma 5.2. Then for some n, for all b, b', we have $|R(b, b')| \leq n$. For any $b, b', b'' \in Q$ it follows from (ii) that $R(b, b') \subseteq R(b, b'') \cup R(b', b'')$, and from (iii) that $R(a, b) = R(b, a)$ and $R(a, a) = \emptyset$.

Let S_0, T be as in Lemma 5.2. By the last statement of this lemma, S_0 is $\mathrm{Aut}(M)$-invariant, and $T(b)$ is $\mathrm{Aut}(M/b)$-invariant. Each element q of the finite set $S_0 \cup T(b)$ thus contains a formula D_b^q defined over b, and of Δ-rank $\mathrm{rk}_\Delta(q) = r$. Taking the disjunction of these formulas, we obtain, for each b with $S_0 \cup T(b) \neq \emptyset$, a formula D_b of Δ-rank r, such that D_b lies in each $q \in S_0 \cup T(b)$. Thus $D_b \vee D_{b'}$ lies in each $q \in S_0 \cup T(b) \cup T(b') \supseteq R(b, b')$. Since $R(b, b')$ contains all Δ-types consistent with $X(b, b')$, it follows that $X'(b, b') := X(b, b') \setminus \{x : D_b(x) \vee D_{b'}(x)\}$ has Δ-rank $< r$. By compactness, we can replace $\{D_b\}$ by a uniformly definable family $\{D_b' : b \in Q\}$ with the same property. For each b, let $D'(b) := \{x : D_b'(x) \text{ holds}\}$.

Now by induction, there exists a uniformly definable $D''(b)$ of Δ-rank $< r$, such that for all $a, b \in Q$ we have $X'(a, b) \subseteq D''(a) \cup D''(b)$.

Let $D(a) = D'(a) \cup D''(a)$; then clearly (i'), (ii') hold. $\qquad\square$

CHAPTER 6

STRONG CODES FOR GERMS

In this chapter we find variants in our context of another well-known result for stable theories: that given the germ of a function on a definable type, one can define from that germ a particular function having the same germ. This phenomenon is important for example in group configuration arguments in stability theory: see for example Ch. 5, Definition 1.3 and the remarks following it in [40]. Versions of this were also important in [12], especially in Section 3.3, where it is shown that in ACVF, the germ of a function on the generic type of a closed ball has a 'strong code' in the sense of the next definition. The main result is Theorem 6.3, which gives strong codes for functions on stably dominated types; the proof uses Lemma 5.1. We also obtain strong codes in Proposition 6.7, where stable domination is replaced by an assumption on the range of the function and the condition '(BS)'. The chapter concludes with some applications of strong codes, such as the useful transitivity result Proposition 6.11.

In this section tuples are finite except where indicated, and types are in finitely many variables.

Definition 6.1. Let p be a C-definable type over $\mathcal{U}$. Let $\varphi(x, y, b)$ be a formula defining a function $f_b(x)$ whose domain contains all realisations of p. The *germ of f_b on p*, or *p-germ of f_b*, is the equivalence class of b under the equivalence relation $\sim$, where $b \sim b'$ if the formula $f_b(x) = f_{b'}(x)$ is in p. Equivalently, $b \sim b'$ if and only if for any $a \models p|Cbb'$, $f_b(a) = f_{b'}(a)$. As p is C-definable, $\sim$ is also C-definable, and the germ of f_b on p is a definable object. A code e for the germ $b/\sim$ of f_b on p is *strong* over C if there is a Ce-definable function g such that the formula $f_b(x) = g(x)$ is in p. Equivalently, the code e is strong if for any $a \models p|Cb$, $f_b(a) \in \mathrm{dcl}(Cea)$.

If p is not definable, but is an $\mathrm{Aut}(\mathcal{U}/C)$-invariant type over $\mathcal{U}$, we still sometimes say that definable functions f and g have the same germ on p. This means that for $a \models p|C^\ulcorner f^\urcorner {}^\ulcorner g^\urcorner$, the formula $f(a) = g(a)$ lies in p. This gives an equivalence relation on any definable family of functions, but in general not a C-definable one.

59

Lemma 6.2. *Suppose that p is a C-definable type over $\mathcal{U}$, and $f = f_b$ is a definable function on the set of realisations of $p|Cb$. Let e be a code for the germ of f on p. Suppose that for any $b' \equiv_{Ce} b$, and any a with $a \models p|Cb$ and $a \models p|Cb'$, we have $f_b(a) = f_{b'}(a)$. Then e is a strong code for the p-germ of f (over C).*

Proof. There is a well-defined function F on p, having the same p-germ as f, and defined by putting $F(a) = f_{b'}(a)$ for any $b' \equiv_{Ce} b$ with $a \models p|Cb'$. The function F is $\mathrm{Aut}(\mathcal{U}/Ce)$-invariant, so there is a formula $\varphi(x, y)$ over Ce such that $\varphi(a, F(a))$ for all $a \models p$, and so that φ determines a function on realisations of p. Adjusting φ, we may suppose it is the graph of a function whose domain contains all realisations of $p|Ce$. The function defined by φ is now Ce-definable, and has the same p-germ as f, as required.　　□

From the lemma we see easily that in a stable theory, strong codes exist. For in the notation of the lemma, suppose $a \models p|Cb$ and $a \models p|Cb'$. Pick $b'' \equiv_{Ce} b$ with $b'' \underset{Ce}{\downarrow} abb'$. Then $f_{b''}$ has the same germ on p as f_b and $f_{b'}$. By properties of non-forking, $a \models p|Cbb''$ and $a \models p|Cb'b''$, so $f_b(a) = f_{b''}(a) = f_{b'}(a)$.

Theorem 6.3. *Let p_0 be a stably dominated type over C with an extension p over $C' := \mathrm{acl}(C)$, and let f be a definable function whose domain contains the set of realisations of $p|\mathcal{U}$. Let $a \models p|\mathcal{U}$ and $C'' := \mathrm{acl}(C) \cap \mathrm{dcl}(Ca)$.*
　(i) *The code for the germ of f on p is strong over C''.*
　(ii) *Suppose that $f(a) \in \mathrm{St}_{Ca}$. Then the code for the p-germ of f is in St_C.*

Proof. (i) We shall work over C'', since by Remark 3.30, $p|C''$ is stably dominated, and by Remark 3.14, p is C''-definable. For convenience suppose $C = C''$, and write p for $p|C''$.

Claim 1. We may assume that p is a type consisting of elements in St_C.

Proof of Claim. For this, we may write the type p as $p(x, y)$, so that if $p(a', a)$ then a' is an enumeration of $\mathrm{St}_C(a)$ (so in general is infinite). Write p_x for the restriction of p to the x-variables. Note that the notion of *germ* of a function on a definable type with infinitely many variables also makes sense.

Write $f = f_b = f(b, x, y)$, where b is the parameter defining f (over C). Put $b \sim b'$ if f_b, $f_{b'}$ have the same p-germ. Then $b \sim b'$ if and only if

　for all $c \models p_x|Cbb'$, for all d realising $p(c, y)|C, f(b, c, d) = f(b', c, d)$.

For by stable domination, $cd \models p|Cbb'$ if and only if $c \models p_x|Cbb'$ and $p(c, d)$ holds. There is a definable function $F_{bc} = F_{bc}(y)$ which agrees with $f(b, c, y)$ on realisations of $p(c, y)$. Since F_{bc} is bc_0-definable for some finite subtuple c_0 of c, we may suppose now that c (and so x) has finite length. Let F_b be the function defined on realisations of $p_x|Cb$, such that $F_b(x) = \ulcorner F_{bx} \urcorner$. Now $b \sim b'$ if and only if the functions $F_b(x)$ and $F_{b'}(x)$ have the same p_x-germ. Furthermore, if the code of $b/\sim$, regarded as the germ of F_b on p_x is strong over C, then it is also strong over C when regarded as the germ of f_b on p. This yields the claim.

Write $f = f_b$, and let Σ be the sort of the type p. We may suppose that f_b is total, by giving it some formal value when undefined. Let e be a code for the p-germ of f and $E := \mathrm{dcl}(Ce)$. Let Q be the set of conjugates of b over E, and P be the set of realisations of $p|E$. For $b, b' \models Q$, let

$$X_0(b, b') := \{x \in \Sigma : f_b(x) \neq f_{b'}(x)\}.$$

Clearly: $X_0(b, b') \subseteq X_0(b, b'') \cup X_0(b', b'')$ for any $b, b', b'' \in Q$.

As St_C is stably embedded, there is a finite set Δ of formulas $\varphi(x, y)$ such that for any $b, b' \in Q$ the set $X_0(b, b')$ is equal to $\varphi(x, d)$ for some tuple $d = d(b, b')$ from St_C and some $\varphi(x, y) \in \Delta$.

Let ρ be the Δ-rank of p; there exists in p a formula $D(x)$ over C equivalent to a Δ-formula of Δ-rank ρ and multiplicity 1. For any Δ-formula $\psi(x)$, we have $\psi(x) \in p|\mathcal{U}$ iff $\mathrm{rk}_\Delta(\psi \wedge D) = \rho$. Put

$$X(b, b') = X_0(b, b') \cap D$$

Observe that if $a \models p|Cbb'$ then $a \notin X(b, b')$, since f_b and $f_{b'}$ have the same p-germ. Hence $\mathrm{rk}_\Delta(X(b, b')) < \rho$.

By Lemma 5.4, uniformly in $b \in Q$ there is a definable set $X(b)$ of Δ-rank $< \rho$ such that for all $b, b' \in Q$, $X(b, b') \subseteq X(b) \cup X(b')$.

It follows that $f_b, f_{b'}$ agree away from $X(b) \cup X(b')$; in particular they agree on any c such that $c \models p|Cb$ and $c \models p|Cb'$. By Lemma 6.2, there is an E-definable function on P with the same germ on P as f_b.

(ii) Again, we may assume $C = \mathrm{acl}(C)$. Let e be a code for the p-germ of f. Under the assumption that $f(a) \in \mathrm{St}_{Ca}$ for all $a \in P$, we must show that $e \in \mathrm{St}_C$. By replacing f by the function defined from e with the same germ, we may suppose that $f = f_e$, and is Ce-definable.

We first need the following claim.

CLAIM 2. There is a natural number n and elements $a_1, \ldots, a_n$ of P such that $e \in \mathrm{dcl}(Ca_1 \ldots a_n f(a_1) \ldots f(a_n))$.

PROOF OF CLAIM. Let $\kappa = \mathrm{wt}(\mathrm{tp}(a^{\mathrm{st}}/C))$, which is bounded in terms of $|T|$ (see Section 2.3). Choose an indiscernible sequence $(a_i : i < \kappa^+)$ of realisations of p, with $a_\lambda \models p|C(a_\mu : \mu < \lambda)$. By Proposition 3.16, $(a_i^{\mathrm{st}} : i < \kappa^+)$ is a Morley sequence over C. Put $d_i = f(a_i)$. Then $e \in \mathrm{dcl}(C, a_i d_i : i < \kappa^+)$. For suppose e' is conjugate to e over $\{C, a_i d_i : i < \kappa^+\}$. Then for each $i < \kappa^+$ we have $f(a_i) = f_{e'}(a_i)$ (where $f_{e'}$ is the function defined from e' in the same way that f is defined from e). By the weight assumption, there is $i < \kappa^+$ with $a_i^{\mathrm{st}} \downarrow_C (ee')^{\mathrm{st}}$. Hence, for any $a \in P$ with $a \downarrow_C^d ee'$, we have $f(a) = f_{e'}(a)$. As e is the code for the germ of f, $e = e'$.

In particular, there are $i_1, \ldots, i_n$ and a C-definable function h with $e = h(a_{i_1}, \ldots, a_{i_n}, d_{i_1}, \ldots, d_{i_n})$.

Since $f(a) \in \mathrm{St}_{Ca}$ for every $a \in P$, there is a C-definable relation $R(x, y)$ so that for each j, $R(a_{i_j}, y)$ defines a stable, stably embedded set which contains d_{i_j}. As p is a definable type, the following defines a first-order formula $S(z)$ over C (where $d_p u \varphi(u)$ means that $\varphi(u)$ holds for $u \models p|\mathcal{U}$):

$$d_p u_1 \dots d_p u_n \exists y_1 \dots \exists y_n \left(\bigwedge_{j=1}^n R(u_j, y_j) \wedge z = h(u_1, \dots, u_n, y_1, \dots, y_n) \right).$$

Certainly $S(e)$ holds, and it remains to check that $S(y)$ is stable and stably embedded. Now S can also be defined, over $Ca_{i_1} \dots a_{i_n}$, by the formula

$$\exists y_1 \dots \exists y_n \left(\bigwedge_{j=1}^n R(a_{i_j}, y_j) \wedge z = h(a_{i_1}, \dots, a_{i_n}, y_1, \dots, y_n) \right).$$

Thus S is internal over the stable, stably embedded set $\bigcup_{j=1}^n R(a_{i_j}, y_j)$ and hence is itself stable and stably embedded. □

DEFINITION 6.4. Let $A = \mathrm{dcl}(Ca)$, and let κ be a cardinal. We say that A^{st} is κ-generated over C if there is a set $Y \subseteq A^{\mathrm{st}}$ such that $|Y| \leq \kappa$ and $A^{\mathrm{st}} \subseteq \mathrm{dcl}(CY)$.

We now work towards a variant of Theorem 6.3 (ii) for arbitrary definable types, under a certain additional assumption (BS) restricting the growth of $\mathrm{St}_C(a)$ as C grows. It is possible that (BS) follows from metastability over an o-minimal Γ; we show at all events later that it holds for ACVF.

LEMMA 6.5. Let A^{st} be any definably closed subset of St_C. Let $\kappa \geq |T|$. Then A^{st} is κ-generated over C if and only if there is no strictly ascending chain of length κ^+ of sets $B = \mathrm{dcl}(B) \cap \mathrm{St}_C$ between C and A^{st}.

PROOF. For the forward direction, assume A^{st} is κ-generated over C by the set Y, let $\lambda = \kappa^+$ and suppose there is a chain $(A_\alpha: \alpha < \lambda)$ of subsets of A^{st} ordered by inclusion, each A_α definably closed in St_C, with $A_0 = C$. By stability, $\mathrm{tp}(Y/\bigcup_i A_i)$ is definable, hence using Corollary 2.18 it is definable over A_μ for some $\mu < \lambda$. Consider $c \in A_{\mu+1}$. Then $c = f(y)$ for some $y \in Y$ and C-definable function f. As $\mathrm{tp}(Y/CA_{\mu+1})$ is A_μ-definable, $\{x \in A_{\mu+1}: f(y) = x\}$ is definable over A_μ and is a singleton. Since A_μ is definably closed, $c \in A_\mu$ and hence the chain stabilises.

For the other direction, suppose A^{st} is not boundedly generated and take $\lambda = (2^{|T|})^+$. Construct a sequence $(c_\alpha: \alpha < \lambda)$ of elements of A^{st} with $c_\alpha \notin \mathrm{dcl}(Cc_\beta: \beta < \alpha)$, for each α. Then for each $\mu < \lambda$, put $A_\mu := \mathrm{dcl}(Cc_\alpha: \alpha < \mu)$. This is a strictly ascending chain of length λ of subsets of St_C. □

DEFINITION 6.6. We say that the theory T has the *bounded stabilising property* (BS) if, for every C and every $A = \mathrm{dcl}(Ca)$, where a is a finite tuple, there is no strictly ascending chain of length $(2^{|T|})^+$ of sets $B = \mathrm{dcl}(B) \cap \mathrm{St}_C$ between C and A^{st}.

We will see in Proposition 9.7 that property (BS) holds in ACVF.

PROPOSITION 6.7. *Assume T has* (BS), *and let p be a C-definable type over $\mathcal{U}$. Let f be a definable function on P, the set of realisations of $p|C$, and suppose that $f(a) \in \mathrm{St}_{Ca}$ for all $a \in P$. Then*

(i) *the code e for the germ of f on p (over C) is strong, and*
(ii) $e \in \mathrm{St}_C$.

PROOF. (i) Suppose that f is b-definable. For any ordered set I and any set B containing b, we can find an indiscernible sequence $\{a_i : i \in I\}$ over CB, with $a_i \models p|C \cup \{Ba_j : j < i\}$ for any $i \in I$. Now let $A := C \cup \{a_i : i \in I\}$. For any $i \in I$ let $f(a_i) = d_i$, and for any $I' \subseteq I$ let $D_{I'} := \mathrm{dcl}(A \cup \{d_i : i \in I'\})$. Since by assumption, $f(a_i) \in \mathrm{St}_{Ca_i}$, $D_{I'} \subseteq \mathrm{St}_A(B)$. By choosing I large, we may find a strictly increasing chain $(I_\alpha : \alpha < (2^{|T|})^+)$ of initial subintervals of I. Then by (BS) (applied within St_A)), the sequence $(D_{I_\alpha} : \alpha < (2^{|T|})^+)$ is eventually constant, so there is $i \in I$ so that $d_i \in \mathrm{dcl}(Ad_j : j < i)$. Choosing I to have no greatest or least element, it follows by indiscernibility that this holds for all $i \in I$. By a similar argument we may also suppose that $d_i \in \mathrm{dcl}(Ad_j : j > i)$. Thus, we have the following, for some $n \in \omega$ and any $i_1 < \cdots < i_n$ from I:

$$d_{i_n} \in \mathrm{dcl}(Ca_{i_1} \ldots a_{i_n} d_{i_1} \ldots d_{i_{n-1}}) \tag{6.1}$$
$$d_{i_1} \in \mathrm{dcl}(CAd_{i_2} \ldots d_{i_n}). \tag{6.2}$$

To see (6.1) above, certainly there are m, n such that for any $j_1, \ldots, j_m$ with $i_1 < \cdots < i_n < j_1 < \cdots < j_m$, $d_{i_n} \in \mathrm{dcl}(Ca_{i_1} \ldots a_{i_n} d_{i_1} \ldots d_{i_{n-1}} a_{j_1} \ldots a_{j_m})$. Suppose $\mathrm{tp}(d'_{i_n}/Ca_{i_1} \ldots a_{i_n} d_{i_1} \ldots d_{i_{n-1}}) = \mathrm{tp}(d_{i_n}/Ca_{i_1} \ldots a_{i_n} d_{i_1} \ldots d_{i_{n-1}})$. Choose $a_{j_1}, \ldots, a_{j_m}$ so that for each k, $a_{j_k} \models p|Ca_{i_1} \ldots a_{i_n} d_{i_1} \ldots d_{i_n} d'_{i_n} a_{j_1} \ldots a_{j_{k-1}}$. Then

$$\mathrm{tp}(d_{i_n}/Ca_{i_1} \ldots a_{i_n} d_{i_1} \ldots d_{i_{n-1}} a_{j_1} \ldots a_{j_m})$$
$$= \mathrm{tp}(d'_{i_n}/Ca_{i_1} \ldots a_{i_n} d_{i_1} \ldots d_{i_{n-1}} a_{j_1} \ldots a_{j_m}).$$

This forces $d_{i_n} = d'_{i_n}$.

We can now show that if f' is a definable function conjugate to f over C and has the same germ on p, then for any a, if $a \models p|C \ulcorner f \urcorner$ and $a \models p|C \ulcorner f' \urcorner$ then $f(a) = f'(a)$. For argue as above with $B = \mathrm{dcl}(C \ulcorner f \urcorner \ulcorner f' \urcorner a)$. Let a^* be an enumeration of A. Then $\ulcorner f \urcorner aa^* \equiv_C \ulcorner f' \urcorner aa^*$, and in particular

$$aa^* f(a) f(a_{i_2}) \ldots f(a_{i_n}) \equiv_C aa^* f'(a) f'(a_{i_2}) \ldots f'(a_{i_n}).$$

As f, f' have the same p-germ, $f(a_{i_j}) = f'(a_{i_j}) = d_j$, say, for each $j = 2, \ldots, n$. By (6.2) above (applied over Cb in place of B), we have $f(a), f'(a) \in$

$\text{dcl}(Caa^*d_2 \ldots d_n)$. Thus $f(a) = f'(a)$. It follows by Lemma 6.2 that e is a strong code for the germ of f over C.

(ii) It suffices to prove (under the present hypotheses) Claim 2 from the proof of Theorem 6.3, since then that proof can be mimicked. Notice that (6.1) implies that any automorphism fixing $Ca_{i_1} \ldots a_{i_n}, d_{i_1}, \ldots d_{i_{n-1}}$ also fixes $f(a_{i_n})$. It follows that $e \in \text{dcl}(Ca_{i_1} \ldots a_{i_{n-1}} d_{i_1} \ldots d_{i_{n-1}})$. This gives Claim 2. $\square$

REMARK 6.8. The proof of (i) yields also the following. Assume (BS), and let p be an $\text{Aut}(\mathcal{U}/C)$-invariant type, not necessarily definable. Suppose that f is a CB-definable function such that if $a \models p|CB$ then $f(a) \in \text{St}_{Ca}$. Suppose also that D is a parameter set containing C, and that for any automorphism σ fixing D pointwise, f and $\sigma(f)$ have the same p-germ. Then there is a D-definable function with the same germ as f on p.

LEMMA 6.9. Let $C \subseteq C'' \subseteq \text{acl}(C)$, and $C \subseteq B$. Then $\text{St}_{C''}(B) = \text{dcl}(\text{St}_C(B)C'')$.

PROOF. Clearly $\text{dcl}(\text{St}_C(B)C'') \subseteq \text{St}_{C''}(B)$.

For the other direction, let $e \in \text{St}_{C''}(B)$. We have $\text{St}_C = \text{St}_{C''} = \text{St}_{\text{acl}(C)}$, so $\text{St}_C''(B) = \text{St}_{C''} \cap \text{dcl}(C''B) = \text{St}_C \cap \text{dcl}(C''B)$. Thus there are finite $c'' \in C''$ and $b \in B$ with $e \in \text{dcl}(c''b)$. Let e' be the (finite) set of conjugates of ec'' over CB. Then $e' \in \text{St}_C(B)$. Also, $\text{tp}(ec''/e') \vdash \text{tp}(ec''/B)$, so $\text{tp}(e/c''e') \vdash \text{tp}(e/c''B)$. It follows that $e \in \text{dcl}(c''e') \subseteq \text{dcl}(c'', \text{St}_C(B)) \subseteq \text{dcl}(\text{St}_C(B)C'')$. $\square$

We now give some applications of the existence of strong codes.

PROPOSITION 6.10. Assume $C \subseteq B$, $\text{tp}(A/C)$ is stably dominated, and $A \underset{C}{\overset{d}{\downarrow}} B$. Let $A^{\text{st}} = \text{St}_C(A)$ Then the following hold.

(i) $\text{tp}(A/B)$ is stably dominated.
(ii) $\text{St}_C(AB) = \text{St}_C(A^{\text{st}}B^{\text{st}})$.
(iii) If $\text{tp}(B/C)$ is stably dominated then $\text{St}_B(A) = \text{dcl}(BA^{\text{st}}) \cap \text{St}_B$.
(iv) If $\text{tp}(B/C)$ is extendable to a definable type, and T satisfies (BS), then $\text{St}_B(A) = \text{dcl}(BA^{\text{st}}) \cap \text{St}_B$.

PROOF. (i) This is just Proposition 4.1.

(ii) Put $C'' := \text{dcl}(Ca) \cap \text{acl}(C)$. Let $d \in \text{St}_C(AB) = \text{dcl}(AB) \cap \text{St}_C$. Then $d = f(a)$ for some $a \in A$ and B-definable function f. By Proposition 3.32 (iii), $\text{tp}(a/C)$ is stably dominated. Let $C'' := \text{acl}(C) \cap \text{dcl}(Ca)$. By Theorem 6.3, the germ of f on the definable extension of $\text{tp}(a/\text{acl}(C))$ is strongly coded with the code e over C'' lying in St_C. Now $e \in \text{St}_{C''}(B)$, so by Lemma 6.9, $e \in \text{dcl}(\text{St}_C(B)C'')$. Since the code is strong, there is a $C''e$-definable function f' with the same germ as f, and hence $d = f'(a)$ is definable over $\text{St}_C(B)C''A$, so is AB^{st}-definable. Now write $d = g(b)$, where $b \in B^{\text{st}}$ and g is an A-definable function. By stable embeddedness, g is A^{st}-definable, hence $d \in \text{dcl}(A^{\text{st}}B^{\text{st}})$.

(iii) Let $d \in \text{St}_B(A) = \text{dcl}(BA) \cap \text{St}_B$. Then $d = g(b)$ for some $b \in B$ and A-definable function g. As $d \in \text{St}_B$, we can expand b to a larger tuple from B to arrange $g(b) \in \text{St}_{Cb}$. By Theorem 6.3 (with the roles of A and B reversed) the code e over $\text{dcl}(Cb) \cap \text{acl}(C)$ for the germ over $C^* := \text{acl}(C) \cap \text{dcl}(Cb)$ of g on the definable extension of $\text{tp}(b/\text{acl}(C))$ is strong. Hence $e \in \text{St}_C$, by Theorem 6.3 (ii). Thus, arguing as in (ii), there is an $\text{St}_C(A)C^*$-definable function g' with $g'(b) = d$, that is, d is definable from $A^{\text{st}}b$.

(iv) The proof is the same as (iii), using Proposition 6.7 instead of Theorem 6.3. □

PROPOSITION 6.11. *Suppose* $\text{tp}(A/C)$ *and* $\text{tp}(B/CA)$ *are stably dominated. Then* $\text{tp}(AB/C)$ *is stably dominated.*

PROOF. By Corollary 3.31 (iii) and Proposition 4.1, we may assume that $C = \text{acl}(C)$. Throughout this argument, for any set X we will write $X^{\text{st}} := \text{dcl}(CX) \cap \text{St}_C$ and $X^* := \text{dcl}(CAX) \cap \text{St}_{CA}$. Notice that $B^* \cap \text{St}_C = (AB)^{\text{st}} = (B^*)^{\text{st}}$. Suppose $(AB)^{\text{st}} \mathop{\smile\hskip-0.8em\lower0.3ex\hbox{\vrule height0.6em}}_C D^{\text{st}}$. We need to show that $\text{tp}(D/C(AB)^{\text{st}}) \vdash \text{tp}(D/CAB)$.

As in Proposition 6.10 we can show that $D^* = \text{dcl}(CAD^{\text{st}}) \cap \text{St}_{CA}$. For consider $d^* \in D^*$. There are $a \in A$ and a CD-definable function h with $h(a) = d^* \in \text{St}_{Ca}$. By Theorem 6.3, the germ (over $C'' := \text{acl}(C) \cap \text{dcl}(Ca)$) of h on the definable extension of $\text{tp}(a/\text{acl}(C))$ is strongly coded in St_C. Hence, as in Proposition 6.10 (ii), there is a function H definable over $\text{St}_C(D)C''$ with $H(a) = d^*$. Thus $d^* \in \text{dcl}(CAD^{\text{st}})$.

By Remark 3.7 over St_{CA}, we know that $\text{tp}(D/CAD^*) \vdash \text{tp}(D/CAD^*B^*)$, and hence by the above paragraph, that $\text{tp}(D/CAD^{\text{st}}) \vdash \text{tp}(D/CAD^{\text{st}}B^*)$. Rewriting this as $\text{tp}(AB^*/CD^{\text{st}}) \vdash \text{tp}(AB^*/CD)$ and recalling the hypothesis that $(AB)^{\text{st}} \mathop{\smile\hskip-0.8em\lower0.3ex\hbox{\vrule height0.6em}}_C D^{\text{st}}$, we have

$$AB^* \mathop{\smile\hskip-0.8em\lower0.3ex\hbox{\vrule height0.6em}}_C^d D. \tag{6.3}$$

This implies that $\text{tp}(D/C(AB^*)^{\text{st}}) \vdash \text{tp}(D/CAB^*)$; that is,

$$\text{tp}(D/C(AB)^{\text{st}}) \vdash \text{tp}(D/CAB^*). \tag{6.4}$$

It follows from (3) that $B^* \mathop{\smile\hskip-0.8em\lower0.3ex\hbox{\vrule height0.6em}}_{CA} D^*$ (in St_{CA}). For there is certainly some B' with $\text{tp}(B'/CA) = \text{tp}(B/CA)$ for which this is true. Intersecting with St_C and applying transitivity (as $A^{\text{st}} \mathop{\smile\hskip-0.8em\lower0.3ex\hbox{\vrule height0.6em}}_C D^{\text{st}}$), we get $(AB')^{\text{st}} \mathop{\smile\hskip-0.8em\lower0.3ex\hbox{\vrule height0.6em}}_C D^{\text{st}}$. Since we also know $(AB)^{\text{st}} \mathop{\smile\hskip-0.8em\lower0.3ex\hbox{\vrule height0.6em}}_C D^{\text{st}}$, it follows that $\text{tp}(D^{\text{st}}(AB)^{\text{st}}/C) = \text{tp}(D^{\text{st}}(AB')^{\text{st}}/C)$. By (3), $\text{tp}(AB^*/CD^{\text{st}}) \vdash \text{tp}(AB^*/CD)$. Hence, as $AB \equiv_{CD^{\text{st}}} AB'$, we have $(AB)^* \equiv_{CD^{\text{st}}} (AB')^*$. Clearly $AB^* = (AB)^*$ and $A(B')^* = (AB')^*$, so $AB^* \equiv_{CD} A(B')^*$. Hence, as $D^* \mathop{\smile\hskip-0.8em\lower0.3ex\hbox{\vrule height0.6em}}_{CA} (B')^*$, we have $D^* \mathop{\smile\hskip-0.8em\lower0.3ex\hbox{\vrule height0.6em}}_{CA} B^*$.

By stable domination of $\text{tp}(B/AC)$ it follows that

$$\text{tp}(D/CAB^*) \vdash \text{tp}(D/CAB).$$

Together with (4), this gives the required result. □

COROLLARY 6.12. *Suppose* $\mathrm{tp}(A/C)$ *is stably dominated. Then* $\mathrm{tp}(\mathrm{acl}(CA)/C)$ *is stably dominated.*

PROOF. This follows from Proposition 6.11, as $\mathrm{tp}(\mathrm{acl}(CA)/CA)$ is clearly stably dominated. ☐

We close this part with an illustration of stable domination and strong germs in the context of groups; cf. [15].

Let G be a definable group, and p a definable type over $\mathcal{U}$ of elements of G (i.e., containing the formula $x \in G$.)

For $a \in G(\mathcal{U})$, we define the *translate* $^a p$ to be $\mathrm{tp}(ag/\mathcal{U})$, where $g \models p|\mathcal{U}$. This gives an action of $G(\mathcal{U})$ on the definable types.

We are interested in translation invariant types. In this case, since $^g p = p$ for $g \in G$, it follows that any element g of G is a product of two elements of p.

THEOREM 6.13. *Let* G *be a definable group,* p *a stably dominated definable type of* G. *Assume* p *is translation invariant. Then there exist definable stable groups* $\mathfrak{g}_i$, *and definable homomorphisms* $g_i : G \to \mathfrak{g}_i$, *such that* p *is stably dominated via* $g = (g_i : i \in I)$.

PROOF. Let $\theta(a)$ enumerate $\mathrm{St}_C(a)$. Then $p|C$ is stably dominated via θ. Consider the map f_a defined by:

$$f_a(b) = \theta(ab).$$

The p-germ is strong, and is in $\mathrm{St}_C(a)$, so it factors through $\theta(a)$: $f_a = f'_{\theta a}$. By stable embeddedness, it factors through $\theta(b)$ too: let $c = f_a(b) = f'_{\theta a}(b)$. Since $c \in \mathrm{St}_C$, $\mathrm{tp}(\theta(a), c/C, \theta(b)) \vdash \mathrm{tp}(\theta(a), c/C, b)$. Thus $c \in \mathrm{dcl}(\theta(a), \theta(b))$; i.e., $\theta(ab) = c = F(\theta(a), \theta(b))$ for $a \models p$, $b \models p|Ca$. Associativity of the group operation on G immediately gives associativity of F on independent triples of realizations of $q = \mathrm{tp}(c/C)$, within the stable structure St_C. Hence by the group chunk theorem of [16] (alternatively see [15], or Poizat's book [42]), there exists an inverse limit system of stable groups $\mathfrak{g}_i$ with inverse limit $\mathfrak{g}$, such that $\theta(a)$ is a generic element of $\mathfrak{g}$. Now θ is generically a homomorphism. If a, b, c, d realize p and $ab = cd$ then $\theta(a)\theta(b) = \theta(c)\theta(d)$. (Let $e \models p|C(a, b, c, d)$; then $abe = cde$, and $\theta(abe) = \theta(a)\theta(be) = \theta(a)\theta(b)\theta(e)$; and similarly for cde.) Using the fact that any element of G is a product of two generics, θ extends uniquely to a homomorphism $G \to \mathfrak{g}$. ☐

PART 2

INDEPENDENCE IN ACVF

CHAPTER 7

SOME BACKGROUND ON ALGEBRAICALLY CLOSED
VALUED FIELDS

In this chapter we give a little background on the model theory of valued
fields, emphasising algebraically closed valued fields. This monograph depends heavily on results and methods from [12]. We shall summarise both
the main results from [12] which we use, and some of the methods developed
there which we shall exploit.

7.1. Background on valued fields

A valued field consists of a field K together with a homomorphism $| - |$
from its multiplicative group to an ordered abelian group Γ, which satisfies the
ultrametric inequality. We shall follow here the notation of [12], and view the
value group $(\Gamma, <, ., 1)$ multiplicatively, with identity 1. Abusing notation, we
shall usually suppose that it contains an additional formal element 0. So $0 < \gamma$
for all $\gamma \in \Gamma$, and the axioms for a valuation are as follows (with $x, y \in K$):
(i) $|xy| = |x|.|y|$;
(ii) $|x + y| \leq \text{Max}\{|x|, |y|\}$;
(iii) $|x| = 0$ if and only if $x = 0$.
The valuation is *non-trivial* if its range properly contains $\{0, 1\}$.

For most arguments in this monograph we find this multiplicative notation
more intuitive than the usual additive one. However, we do occasionally adopt
additive notation (with value map denoted v) during more valuation-theoretic
arguments. For example, Lemma 12.16 and the proof of Proposition 12.15
are written additively, as are 13.4 and 13.6. Also, viewed additively, the value
group Γ of a field is a vector space over $\mathbb{Q}$, and we sometimes write $\text{rk}_{\mathbb{Q}}(\Gamma)$ for
its vector space dimension, even when viewing it multiplicatively.

If the value map $| - |: K \to \Gamma$ is surjective, we say that K has *value group* Γ
(though formally the group has domain $\Gamma \setminus \{0\}$). Often the value map is
implicit, and we just refer to the valued field (K, Γ). The *valuation ring* of K
is $R := \{x \in K : |x| \leq 1\}$. This is a local ring with unique maximal ideal

$\mathcal{M} = \{x : |x| < 1\}$. The *residue field* is $k := R/\mathcal{M}$, and there is a natural map res: $R \to k$. Later, when talking about ACVF, we shall slightly adjust this notation, viewing $K, \Gamma, R, \mathcal{M}, k$ as definable objects in a large saturated model of ACVF. If $\gamma \in \Gamma$ we often use the notation $\gamma R = \{x \in K : |x| \leq \gamma\}$ and $\gamma \mathcal{M} = \{x \in K : |x| < \gamma\}$.

In many texts (for example Ribenboim [44]), part of the definition of *valued field* requires that the value group (written additively) is archimedean, that is, embeds in $(\mathbb{R}, +)$. Such sources refer to our more general notion as a *Krull valuation*. Since we work with saturated models, for which the value group will be non-archimedean, the more general setting is forced on us.

The most familiar valued fields are probably the p-adic fields $\mathbb{Q}_p$. Other examples (written additively) are the fields of rational functions $F(T)$ (F any field): here, if $p(T), q(T) \in F[T]$ are coprime, then $v(p(T)/q(T)) = \deg(p(T)) - \deg(q(T))$. Also, if F is any field, and Γ is any ordered abelian group (written additively), we may form the field of generalised power series $F((T))^{\Gamma}$ consisting of elements $\Sigma_{\gamma \in \Gamma} a_{\gamma} T^{\gamma}$ whose *support* $\{\gamma \in \Gamma : a_{\gamma} \neq 0\}$ is well-ordered. Addition and multiplication are defined as for power series, and we put $v(\Sigma_{\gamma \in \Gamma} a_{\gamma} T^{\gamma}) = \min\{\gamma \in \Gamma : a_{\gamma} \neq 0\}$. The field $F((T))^{\Gamma}$ has value group Γ and residue field F. In the particular case when Γ is isomorphic to $(\mathbb{Z}, +)$ we write $F((T))^{\Gamma}$ as $F((T))$, the field of Laurent series.

Given a valued field (K, Γ) with residue field k, there are three possibilities for the pair (char(K), char(k)): $(0, 0), (p, p)$, and $(0, p)$ (the 'mixed characteristic' case). For $\mathbb{Q}_p$ we have $(0, p)$, and for generalised power series fields $F((T))^{\Gamma}$, the field has the same characteristic as its residue field F.

There are various natural ways to view a valued field model-theoretically, and these mostly give the same universe of interpretable sets. The simplest is to view the object as having two sorts K and Γ, with the value map between them. The same structure can also be parsed just in the field sort, with a unary predicate for the valuation ring, or with a binary predicate interpreted as $|x| \leq |y|$. Indeed, if the valuation ring R is specified, and U is its group of units, then the value group Γ is isomorphic to K^*/U, so the valuation is determined up to an automorphism of the value group. Another option is to view it as a pair $(K, k \cup \{\infty\})$ with a place $\pi : K \to k \cup \{\infty\}$ where $\pi(x) = \mathrm{res}(x)$ if $x \in R$, and $\pi(x) = \infty$ for $x \notin R$. Recall here that a *place* is a map $\pi : K \to F \cup \{\infty\}$ (where K, F are fields) such that $\varphi(x + y) = \varphi(x) + \varphi(y)$ and $\varphi(xy) = \varphi(x)\varphi(y)$ whenever the expressions on the right hand side are defined, and such that $\varphi(1) = 1$. Any surjective place $\pi : K \to F \cup \{\infty\}$ determines the valuation ring (as $\{x \in K : \pi(x) \in F\}$), so determines a valuation on K with residue field F, uniquely up to an automorphism of the value group.

Chevalley's Place Extension Theorem can be stated as: given a valued field (K, Γ) and a field extension $L > K$, it is always possible to extend the valuation to L. The value group and residue field of K will embed canonically in those of L. We say that an extension $K < L$ is *immediate* if K and L have the same value group and the same residue field. The valued field (K, Γ) is *maximally complete* if it has no proper immediate extensions.

Generalised power series fields $F((T))^\Gamma$ are maximally complete, as is $\mathbb{Q}_p$. Every valued field (K, Γ) has an immediate maximally complete extension. The key ingredient in the proof of this is the notion of *pseudo-convergent (p.c.) sequence*: the sequence $(a_\gamma : \gamma < \alpha)$ (α an ordinal) is *pseudo-convergent* if $|a_\nu - a_\mu| < |a_\lambda - a_\mu|$ whenever $\lambda < \mu < \nu$. The element $a \in K$ is a *pseudo-limit* of $(a_\gamma : \gamma < \alpha)$ if there is $\lambda < \alpha$ such that for all $\mu, \nu < \alpha$ with $\lambda < \mu < \nu$, $|a_\mu - a_\nu| = |a - a_\mu|$. The valued field (K, Γ) is maximally complete if and only if every p.c. sequence from it has a pseudo-limit in K. It is straightforward, given a p.c. sequence in (K, Γ) to adjoin a pseudo-limit in an immediate extension, and iteration of this procedure yields a maximally complete immediate extension. A detailed study of maximally complete fields, with criteria for uniqueness of immediate maximally complete extensions, was undertaken by Kaplansky [26, 27].

The valued field (K, Γ) is *Henselian* if for any monic polynomial $f(X) \in R[X]$ and any simple root $\alpha \in K$ of res(f) (the reduction of f modulo $\mathcal{M}$), there is $a \in R$ such that $f(a) = 0$ and res(a) = α. This property is expressible by a first order axiom scheme. Every valued field K has a *henselisation*, K^h. This is an immediate valued field extension of K which is henselian, is an algebraic field extension of K, and has the property that the valuation has a unique extension from K^h to the algebraic closure (denoted K^{alg} in this text) of K. The henselisation is unique up to valued-fields isomorphism over K (and the isomorphism is unique). Any maximally complete valued field is henselian, but the converse is in general false.

7.2. Some model theory of valued fields

The best known results in the model theory of valued fields are the Ax–Kochen/Ershov principles ([1], [2], [3], [10]). This is a body of results which reduce problems in the elementary theory of valued fields to that of the value group and residue field. One version states that if two henselian valued fields of residue characteristic zero have elementarily equivalent value groups and residue fields, then the valued fields themselves are elementarily equivalent. The results and methods also give information, for example model completeness, for Th($\mathbb{Q}_p$). The methods do not handle the general case of residue characteristic p, but yield, for example, that for any sentence φ in a language for valued fields, for sufficiently large p, φ holds in $\mathbb{Q}_p$ if and only if it holds in $\mathbb{F}_p((T))$.

The model completeness for $\mathrm{Th}(\mathbb{Q}_p)$ was extended to a quantifier elimination by Macintyre in [33], in the language of rings extended by predicates P_n (for each $n \geq 2$) interpreted by the set of nth powers. The obvious analogy is the Tarski quantifier-elimination for real closed fields, when the order relation is adjoined to the language of rings.

AKE and quantifier elimination results have been greatly extended by Prestel and Roquette, Basarab, Pas, and F-V. Kuhlmann. The AKE reduction to value group and residue field is picked up, at the level of Grothendieck rings, in the very recent work of Hrushovski and Kazhdan [19].

7.3. Basics of ACVF

Suppose now that K is an algebraically closed field, equipped with a surjective and non-trivial valuation $|-|: K \to \Gamma$. It is immediate that the value group Γ is divisible and that the residue field k is algebraically closed. It follows that the valuation topology is not locally compact.

The easiest examples of algebraically closed valued fields to describe, in characteristics $(0,0)$ and (p, p), are the generalised power series fields $F((T))^\Gamma$ where Γ is a divisible ordered abelian group and F is an algebraically closed field (of characteristic 0 and p respectively). If Γ (written additively) is $(\mathbb{Q}, +)$, and F has characteristic 0, then $F((T^\Gamma))$ is more familiar as the field of *Puiseux series*, that is $\bigcup_{n=1}^\infty F((T^{1/n}))$. Another familiar algebraically closed valued field is $\mathbb{C}_p$, the completion of the algebraic closure of $\mathbb{Q}_p$. It has characteristic 0, residue field $\mathbb{F}_p^{\mathrm{alg}}$, and value group (viewed additively) isomorphic to $(\mathbb{Q}, +)$.

The model theory of algebraically closed valued fields was initiated by Abraham Robinson well before the AKE principles, in [45]. He showed that any complete theory of non-trivially valued algebraically closed fields is determined by the pair $(\mathrm{char}(K), \mathrm{char}(k))$.

Robinson's results were stated in terms of model-completeness, but with a little extra work yield the following. Part (iii) below is proved in [12] (Theorem 2.1.1). The language $\mathcal{L}_{\mathrm{div}}$ is the language $(+, -, ., 0, 1, \mathrm{div})$, where div is the binary predicate of K interpreted by $\mathrm{div}(x, y)$ whenever $|y| \leq |x|$.

THEOREM 7.1. *Let K be an algebraically closed valued field.*

(i) *The theory of K has quantifier elimination in the language $\mathcal{L}_{\mathrm{div}}$.*

(ii) *The theory of K has quantifier elimination in a 2-sorted language with a sort K for the field (equipped with the language of rings), a sort Γ for the value group written multiplicatively (with the language $(<, ., 0)$ with usual conventions for 0), and a value map $|-|: K \to \Gamma$ with $|0| = 0$.*

(iii) *The theory of K has quantifier elimination in a 3-sorted language $\mathcal{L}_{\Gamma k}$ with the sorts and language of (ii) together with a sort k for the residue field,* *with the language of rings, and a map* Res: $K^2 \to k$ *given by putting* Res(x, y) *equal to the residue of* xy^{-1} *(and taking value* $0 \in k$ *if* $|x| > |y|$*).*

A key ingredient in proofs of such results is that if $K < L$ is a finite Galois extension, or if L is the algebraic closure of K, and $|-|$ is a valuation on K, then the valuations on L which extend $|-|$ are all conjugate under Gal(L/K).

There is a partial converse of (i), due to Macintyre, McKenna and van den Dries [34]: any non-trivially valued field whose theory has quantifier elimination in the language $\mathcal{L}_{\mathrm{div}}$ is algebraically closed.

If (K, Γ) is a valued field, then an *open ball of radius* γ in K is a set of the form $B_\gamma(a) := \{x \in K : |x - a| < \gamma\}$, and a closed ball of radius γ has form $B_{\leq\gamma}(a) := \{x \in K : |x - a| \leq \gamma\}$ (where $a \in K$ and $\gamma \in \Gamma$). Here, we allow $\gamma = 0$, so view field elements as closed balls. Of course, in the valuation topology on K, both open balls and closed balls of non-zero radius are clopen. It follows easily from quantifier elimination ((i) above) that if K is algebraically closed, then any parameter-definable subset of K (i.e., one-variable set) is a Boolean combination of balls. Jan Holly's more precise statement, in terms of Swiss cheeses, will be given when we discuss 1-torsors. Quantifier elimination very easily yields the following. It also ensures that algebraically closed valued fields are *C-minimal*: this is a variant of o-minimality introduced in [35] and [13], and extended in [19].

PROPOSITION 7.2. [12, *Proposition* 2.1.3] (i) *The value group* Γ *of* K *is o-minimal in the sense that every* K-*definable subset of* Γ *is a finite union of intervals.*

(ii) *The residue field* k *is strongly minimal in the sense that any* K-*definable subset of* k *is finite or cofinite (uniformly in the parameters).*

(iii) Γ *is stably embedded in* K.

(iv) *If* $A \subset K$ *then the model-theoretic algebraic closure* acl$(A) \cap K$ *of* A *in the field sort* K *is equal to the field-theoretic algebraic closure.*

(v) *If* $S \subset k$ *and* $\alpha \in k$ *and* $\alpha \in$ acl(S) *(in the sense of* K^{eq}*), then* α *is in the field-theoretic algebraic closure of* S *in the sense of* k.

(vi) k *is stably embedded in* K.

7.4. Imaginaries, and the ACVF sorts

It is easily seen that the theory of algebraically closed valued fields does not have elimination of imaginaries just in the field sort, or even when the sorts Γ and k (which are $\emptyset$-definable quotients of subsets of K) are added. When the work in [12] was begun, we had expected to prove elimination of imaginaries when sorts for open and closed balls are added, but this too proved false [12, Proposition 3.5.1]. The main result of [12] was the identification of certain sorts (from T^{eq}) for which algebraically closed valued fields do have elimination of imaginaries.

In addition to the sorts K, k, Γ we shall describe below certain sorts S_n and T_n (for $n \geq 1$). We write ACVF for the theory of algebraically closed fields with a non-trivial valuation in a multisorted language $\mathcal{L}_G$ with the sorts K, Γ, k, S_n, T_n (for $n \geq 1$), and we call these the *geometric sorts*. We denote by $\mathcal{U}$ a large sufficiently saturated model of ACVF in these sorts, so write $s \in \mathcal{U}$ to mean that s is a member of one of these sorts (in the large model). Just occasionally we will consider a type p in say the field sort *over* $\mathcal{U}$, and might say that $a \in K$ realises p; that is, we sometimes regard K, Γ etc. as sorts, and sometimes as the corresponding subsets of the large model $\mathcal{U}$. Here K is the sort for the field with the usual ring language, $\Gamma \setminus \{0\}$ is the value group in the language of multiplicative ordered groups, and the valuation is given as a norm $|-|: K \to \Gamma \cup \{0\}$. As above, the ultrametric inequality has the form $|x + y| \leq \text{Max}\{|x|, |y|\}$, and $|x| = 0$ if and only if $x = 0$. Write R for the valuation ring of K and $\mathcal{M}$ for its maximal ideal. The residue field is $k = R/\mathcal{M}$, endowed with the usual ring language, and the 'residue map' from $K \times K$ to k is defined by $\text{Res}(x, y)$: this is the residue of x/y in k if $|x| \leq |y|$, and is 0 otherwise. For $x \in R$, we denote by $\text{res}(x)$ its residue in k.

The sorts S_n and T_n (for $n \geq 1$) are defined as follows. First, S_n is the collection of all codes for free R-submodules of K^n on n generators (we will also call these *lattices*, or *R-lattices*). Thus, if we identify $\text{GL}_n(K)$ with the set of all ordered bases of the K-vector space K^n, then there is a map $\rho_n: \text{GL}_n(K) \to S_n$ taking each basis to a code for the R-lattice spanned by it; the maps ρ_n are part of our language $\mathcal{L}_G$ (viewing $\text{GL}_n(K)$ as a subset of K^{n^2}). In particular, elements of S_n act as codes for the elements of a $\emptyset$-definable quotient of a subset of K^{n^2}. For $s \in S_n$, we shall often write $\Lambda(s)$ for the lattice coded by s. Observe that R^n is a rank n-lattice, so has a ($\emptyset$-definable) code in S_n. Also, for each $\gamma \in \Gamma \setminus \{0\}$, $\gamma R := B_{\leq \gamma}(0)$ is a rank one lattice, so has a code in S_1; the latter is interdefinable with γ.

For $s \in S_n$, write $\text{red}(s)$ for $\Lambda(s)/\mathcal{M}\Lambda(s)$. This has $\emptyset$-definably the structure of an n-dimensional vector space over k. Let T_n be the set of codes for elements of $\bigcup\{\text{red}(s): s \in S_n\}$; that is, each $t \in T_n$ is a coset for some member of $\text{red}(s)$ for some $s \in S_n$, so is a code for a coset of $\mathcal{M}\Lambda(s)$ in $\Lambda(s)$. For each $n \geq 1$, we have the functions $\tau_n: T_n \to S_n$ defined by $\tau_n(t) = s$ if and only if t codes an element of $\text{red}(s)$. We shall put $\mathcal{S} := \bigcup_{n \geq 1} S_n$ and $\mathcal{T} := \bigcup_{n \geq 1} T_n$.

As shown in Section 2.4 of [12], the sorts S_n and T_n can be described as codes for members of coset spaces of matrix groups. This both helps to give an intuition about them, and also supports the proofs, for example in Chapter 11. We sketch the details.

First, observe that $GL_n(K)$ has a $\emptyset$-definable transitive action on S_n: for each $A \in GL_n(K)$, if $s \in S_n$ and B is a matrix whose columns are a basis for $\Lambda(s)$, then $A(s)$ is a code for the lattice with basis the columns of AB. The stabiliser of the code for R^n is just $GL_n(R)$, the group of $n \times n$ matrices over R which are invertible over R. Thus, S_n can be regarded as a set of codes for the coset space $GL_n(K)/GL_n(R)$. However, we find it more useful to work with the following upper triangular representation.

It is noted in [12, Lemma 2.4.8] that every R-lattice A in K^n has a basis such that the corresponding matrix is upper triangular. Indeed, let $\pi_i: K^n \to K^{n-i}$ be the projection to the last $(n-i)$-coordinates, and let $A_i := \ker(\pi_i)$. Then for each i, $A_i \cong R^i$ and $A_{i+1}/A_i \cong R$ (see the proof of [12, Proposition 2.3.10]). It is possible to choose a basis $(u_1, \ldots, u_n)$ of A such that $(u_1, \ldots, u_i)$ is a basis of the free R-module A_i for each i: choose u_{i+1} so that its image generates A_{i+1}/A_i. The matrix whose i^{th} column is u_i for each i is upper triangular.

Let $B_n(K) \subset GL_n(K)$ be the group of invertible upper triangular matrices over K, and $B_n(R)$ be the corresponding subgroup of $GL_n(R)$ (where inverses are required to be over R). Let $TB(K)$ be the set of triangular bases of K^n, that is, bases $(v_1, \ldots, v_n)$ where $v_i \in K^i \times (0)$ (i.e., the last $n-i$ entries of v_i are zero). An element $a = (v_1, \ldots, v_n) \in TB(K)$ can be identified with an element of $B_n(K)$, with v_i as the ith column. Now $B_n(R)$ acts on $B_n(K) = TB(K)$ on the right. Two elements M, M' of $TB(K)$ generate the same rank n lattice precisely if there is some $N \in GL_n(R)$ with $MN = M'$, and as $M, M' \in B_n(K)$, we must have $N \in GL_n(R) \cap B_n(K) = B_n(R)$. Using the last paragraph, this gives an identification of S_n with the set of orbits of $B_n(R)$ on $TB(K)$. Equivalently, S_n can be identified with the set of (codes for) left cosets of $B_n(R)$ in $B_n(K)$. This is a natural way of regarding S_n as a quotient of a power of K by a $\emptyset$-definable equivalence relation.

We can also treat T_n as a set of codes for a finite union of coset spaces. For each $m = 1, \ldots, n$, let $B_{n,m}(k)$ be the set of elements of $B_n(k)$ whose m^{th} column has a 1 in the mth entry and other entries zero. Let $B_{n,m}(R)$ be the set of matrices in $B_n(R)$ which reduce (coefficientwise) modulo $\mathcal{M}$ to an element of $B_{n,m}(k)$. Also define $B_{n,0}(R) = B_n(R)$. Then $B_{n,m}(k)$ and $B_{n,m}(R)$ are groups. Let $e \in S_n$, and put $V := \text{red}(e)$. We may put $\Lambda(e) = aB_n(R)$ for some $a = (a_1, \ldots, a_n) \in TB(K)$. So $\Lambda(e)$ is the orbit of a under $B_n(R)$, or the left coset $aB_n(R)$ where a is regarded as a member of $B_n(K)$, and $(a_1, \ldots, a_n)$ is a triangular basis of the lattice $\Lambda(e)$. There is a filtration

$$\{0\} = V_0 < V_1 < \cdots < V_{n-1} < V_n$$

of V, where V_i is the k-subspace of $\text{red}(e)$ spanned by $\{\text{red}(a_1), \ldots, \text{red}(a_i)\}$ (here $\text{red}(a_j) = a_j + \mathcal{M}e$). The filtration is canonical, in that if also $\Lambda(e) = a'B_n(R)$ then V_i is spanned by $\{\text{red}(a_1'), \ldots, \text{red}(a_i')\}$ for each i.

Let $\mathrm{TB}(V)$ be the set of triangular bases of V, that is, bases $(v_1, \ldots, v_n)$ where $v_i \in V_i \setminus V_{i-1}$. Now $B_n(k)$ acts sharply transitively on $\mathrm{TB}(V)$ on the right, with the action defined by

$$(v_1, \ldots, v_n)(a_{ij}) = (a_{11}v_1, a_{12}v_1 + a_{22}v_2, \ldots, \Sigma_{i=1}^n a_{in}v_i).$$

For each $i = 0, \ldots, n$, put $O_i(V) = V_i \setminus V_{i-1}$ (so $O_0(V) = \{0\}$). It is easily verified that two elements of $\mathrm{TB}(V)$ are in the same orbit under $B_{n,m}(k)$ precisely if they agree in the mth entry. Thus, $O_m(V)$ (the set of mth entries of triangular bases) can be identified with the left coset space $\mathrm{TB}(V)/B_{n,m}(k)$, and $V \setminus \{0\}$ with $\bigcup_{m=1}^n \mathrm{TB}(V)/B_{n,m}(k)$. Now, if M is the triangular basis $(a_1, \ldots, a_n)$ of the lattice with code $e \in S_n$, put $\mathrm{RED}(M) := (a_1 + \mathcal{M}\Lambda(e), \ldots, a_n + \mathcal{M}\Lambda(e))$, a triangular basis for V.

CLAIM. If $M, M' \in \mathrm{TB}(K)$, then they are $B_{n,m}(R)$-conjugate (i.e., there is $N \in B_{n,m}(R)$ with $MN = M'$) precisely if they generate the same lattice, and their reductions $\mathrm{RED}(M)$ and $\mathrm{RED}(M')$ are $B_{n,m}(k)$-conjugate.

PROOF OF CLAIM. Suppose $MN = M'$, where $N \in B_{n,m}(R)$. Then $MB_n(R) = M'B_n(R)$, so M, M' generate the same lattice. Also, reducing mod $\mathcal{M}$, we have $\mathrm{RED}(M)\,\mathrm{red}(N) = \mathrm{RED}(M')$ where $\mathrm{red}(N) \in B_{n,m}(k)$ is obtained by reducing each entry mod $\mathcal{M}$. Conversely, suppose M, M' generate the same lattice. Then there is $a \in B_n(R)$ such that $MA = M'$. Thus, $\mathrm{RED}(M)\,\mathrm{red}(A) = \mathrm{RED}(M')$. Suppose also there is $B \in B_{n,m}(k)$ with $\mathrm{RED}(M)B = \mathrm{RED}(M')$. Then $\mathrm{red}(A) = B$, so $A \in B_{n,m}(R)$.

By the claim and the paragraph before it, M, M' are $B_{n,m}(R)$-conjugate precisely if they generate the same lattice A, and $\mathrm{RED}(M)$, $\mathrm{RED}(M')$ have the same element of $\mathrm{red}(A)$ in the mth entry. Via the identification of $\mathrm{TB}(K)$ with $B_n(K)$, we now obtain an $\emptyset$-definable map $\varphi \colon \bigcup_{m=0}^n B_n(K)/B_{n,m}(R) \to T_n$. For $m \in \{1, \ldots, n\}$ and $M \in B_n(K)$ let $\varphi(MB_{n,m}(R)) = v + \mathcal{M}A$, where A is the lattice spanned by the columns of M, and v is the mth entry of $\mathrm{RED}(M)$. Also put $\varphi(MB_{n,0}(R)) := \mathcal{M}A$, where A is the lattice spanned by the columns of M. Thus, we may identify T_n with $\bigcup_{m=0}^n B_n(K)/B_{n,m}(R)$.

From now on, we view S_n as a set of codes for members of $B_n(K)/B_n(R)$, and T_n as a set of codes for elements of $\bigcup_{m=0}^n B_n(K)/B_{n,m}(R)$.

Formally, in the language $\mathcal{L}_G$, in addition to the maps $\rho_n \colon \mathrm{GL}_n(K) \to S_n$ and $\tau_n \colon T_n \to S_n$, there are, for each $n \geq 1$ and $m = 0, \ldots, n$, maps $\sigma_{n,m} \colon B_n(K) \to T_n$: for $A \in B_n(K)$, let $\sigma_{n,m}(A)$ be a code for the coset $AB_{n,m}(R)$. As mentioned above, $\mathcal{L}_G$ also has the ring language on K and k, the ordered group language on $\Gamma \setminus \{0\}$, and the value map $|-| \colon K \to \Gamma$ and map $\mathrm{Res} \colon K^2 \to k$. Any completion of ACVF admits elimination of quantifiers in a specific definitional expansion of $\mathcal{L}_G$, as shown in [12, Section 3.1]. We will not need the precise details of the expansion here. The completions of ACVF are determined by the characteristics of K and k. The main theorem of [12] is the following.

THEOREM 7.3. *ACVF admits elimination of imaginaries in the sorts of* $\mathcal{G}$.

We work inside a large, homogeneous, sufficiently saturated model $\mathcal{U}$ of ACVF, in the sorts of $\mathcal{G}$, namely K, Γ, k, S_n, T_n (for $n > 0$). By *substructure of* $\mathcal{U}$, we generally mean 'definably closed subset of $\mathcal{U}$'. We occasionally refer to elements of K^{eq}, but by elimination of imaginaries, these will have codes in $\mathcal{U}$. Sometimes realisations of types over $\mathcal{U}$ are considered, but we will assume that all sets of parameters come from $\mathcal{U}$, are small relative to the size of $\mathcal{U}$, and can contain elements of all the geometric sorts unless specifically excluded. If $C \subset \mathcal{U}$, we write $\Gamma(C) = \text{dcl}(C) \cap \Gamma$ and $k(C) = \text{dcl}(C) \cap k$. If C is a subfield of $\mathcal{U}$, not necessarily algebraically closed, we write Γ_C for the value group of C, and k_C for the residue field. Thus, $\Gamma(C) = \mathbb{Q} \otimes \Gamma_C$. Suppose $C \subseteq A$ are sets, with $A = (a_\alpha : \alpha < \lambda)$. As in Part I, when we refer to $\text{tp}(A/C)$, we mean the type of the infinite tuple listing A, indexed by λ; the particular enumeration is not important, and is often omitted. Likewise, if h is an automorphism of some model then $h(A)$ will denote the tuple $(h(a_\alpha) : \alpha < \lambda)$, and the statement $h(A) = g(A)$ means that the corresponding tuples are equal.

REMARK 7.4. By Lemma 2.2.6(ii) of [12], any closed ball u of non-zero radius has code interdefinable with an element of $S_1 \cup S_2$. If the ball contains 0 then it is already a 1-dimensional lattice; and otherwise, u is $\emptyset$-interdefinable with the code for the R-submodule of K^2 generated by $\{1\} \times u$.

7.5. The sorts internal to the residue field

The following lemma is used repeatedly. Part (i) is Lemma 2.1.7 of [12], and (ii) is an easy adaptation. We emphasise that for each $s \in S_n$, $\Lambda(s)$ is definably R-module isomorphic to R^n, but this isomorphism is in general not canonical, as $\Lambda(s)$ has no canonical basis. The point below is that over any algebraically closed base C *in the field sort*, any C-definable lattice has a C-definable basis. If the valuation on C is trivial, this holds by (iii), and if it is non-trivial, it holds as C is the field sort of a model of ACVF.

LEMMA 7.5. [12, *Lemma 2.1.7*] *Let* C *be an algebraically closed valued field* (*or more generally suppose* $\text{acl}_K(C \cap K) \subseteq C \subseteq \text{acl}(C \cap K)$). *Suppose* $s \in S_n \cap \text{dcl}(C)$

(i) $\Lambda(s)$ *is* C-*definably isomorphic to* R^n, *and there are* $a_1, \ldots, a_n \in C^n$ *which form an* R-*basis for* $\Lambda(s)$.

(ii) $\text{red}(s)$ *is* C-*definably isomorphic to* k^n.

(iii) *If the valuation on* C *is trivial,* $\Lambda(s) = R^n$.

For each $s \in S_n$, $\text{red}(s)$ is a finite-dimensional vector space over k. As the residue field is a stable, stably embedded subset of the structure, so is $\text{red}(s)$. As we have seen in Part I, the stable, stably embedded sets can play an important role for the independence theory of a structure. In an algebraically closed valued field we give the following definition.

DEFINITION 7.6. For any parameter set C, let $\mathrm{VS}_{k,C}$ be the many-sorted structure whose sorts are the k-vector spaces $\mathrm{red}(s)$ where $s \in \mathrm{dcl}(C) \cap S$. Each sort $\mathrm{red}(s)$ is equipped with its k-vector space structure. In addition, $\mathrm{VS}_{k,C}$ has, as its $\emptyset$-definable relations, any C-definable relations on products of the sorts.

In [12] we used the notation Int rather than VS; we have changed the notation to $\mathrm{VS}_{k,C}$ to emphasise that the structure consists of vector spaces, and not *all* of the k-internal sets.

PROPOSITION 7.7. [12, *Proposition* 2.6.5] *For any parameter set* C, $\mathrm{VS}_{k,C}$ *has elimination of imaginaries.*

Clearly, $\mathrm{VS}_{k,C}$ is contained in St_C. By the following proposition from [12], they are essentially the same. Recall that a C-definable set D is k-*internal* if there is finite $F \subset \mathcal{U}$ such that $D \subset \mathrm{dcl}(k \cup F)$. It is clear that if $s \in S_n$ is C-definable, then $\mathrm{red}(s)$ is k-internal; for if B is a basis of $\mathrm{red}(s)$, then $\mathrm{red}(s) \subset \mathrm{dcl}(k \cup B)$.

PROPOSITION 7.8. [12, *Proposition* 3.4.11] *Let* D *be a* C-*definable subset of* K^{eq}. *Then*

(i) D *is* k-*internal if and only if* D *is stable and stably embedded.*
(ii) *If* D *is* k-*internal then* $D \subset \mathrm{dcl}(C \cup \mathrm{VS}_{k,C})$.

REMARK 7.9. It follows by the last proposition and Remark 3.6 that condition $(*_C)$ of Lemma 3.5 holds in ACVF.

In [12, Lemma 2.6.2] several other conditions equivalent to k-internality are given. In particular, we have

LEMMA 7.10. [12, *Lemma* 2.6.2] *Let* D *be a* C-*definable set. Then the following are equivalent.*

(i) D *is* k-*internal.*
(ii) D *is finite or* (*after permutation of coordinates*) *contained in a finite union of sets of the form* $\mathrm{red}(s_1) \times \cdots \times \mathrm{red}(s_m) \times F$, *where* $s_1, \ldots, s_m$ *are* $\mathrm{acl}(C)$-*definable elements of* S *and* F *is a* C-*definable finite set of tuples.*

7.6. Unary sets, 1-torsors, and generic 1-types

In strongly minimal and o-minimal contexts, one often argues by induction on dimension, fibering an n-dimensional set over an $n-1$-dimensional set with 1-dimensional fibers, thus reducing many questions to the one-dimensional case over parameters. This can also be done for definable subsets of K^n, when

$K \models ACVF$, but is less clear for the lattice sorts S_n and for T_n. It turns out however that a good substitute exists, and we proceed to describe it.

In particular, we will use this process to define *sequential independence*. As noted earlier, one variable definable sets in the field sorts are finite unions of balls. This yields a natural notion of independence: if $C \subseteq B$ and $a \in K$, then a is independent from B over C if any B-definable ball containing a contains a C-definable ball (not necessarily properly) containing a. We shall develop the theory in the more general setting of 'unary sets', to obtain a form of independence for the S_n and T_n sorts.

First, recall that a *torsor* U of an R-module A is a set equipped with a regular i.e., sharply 1-transitive) action of A on U. A *subtorsor* is a subset of U of the form $u + B$, where B is an R-submodule of A.

For $\gamma \in \Gamma$, let $\gamma R := \{x \in K : |x| \le \gamma\}$ and $\gamma \mathcal{M} = \{x \in K : |x| < \gamma\}$. These are both R-submodules of K. A *definable 1-module* is an R-module in K^{eq} which is definably isomorphic to $A/\gamma B$, where A is one of K, R or $\mathcal{M}$, B is one of R or $\mathcal{M}$ and $\gamma \in \Gamma$ with $0 \le \gamma \le 1$. The 1-module is *open* if A is K or $\mathcal{M}$, and *closed* if A is R. A *definable 1-torsor* is a set equipped with a definable regular action of a definable 1-module on it, and an ∞-*definable* 1-torsor is an intersection of a chain (ordered under inclusion) of definable subtorsors of a definable 1-torsor. Typically, (∞-) definable 1-torsors arise as cosets of (∞-) definable 1-modules, in a larger 1-module. We will sometimes call a definable 1-torsor a *ball* if $\gamma = 0$. A *1-torsor* is either a definable 1-torsor or an ∞-definable 1-torsor, and we call it a *C-1-torsor* if the parameters used to define it come from C; we do not here require that the isomorphism to $R/\gamma \mathcal{M}$, etc, is C-definable. Finally, a *C-unary set* is a C-1-torsor or an interval $[0, \alpha)$ in Γ. A *unary type* over C is the type of an element of a C-unary set. The most natural examples of unary sets are just balls.

LEMMA 7.11. *Suppose a lies in a C-unary set U. Then $\mathrm{rk}_{\mathbb{Q}}(\Gamma(Ca)/\Gamma(C)) + \mathrm{trdeg}(k(Ca)/k(C)) \le 1$.*

PROOF. Let M be any model containing C. Suppose the lemma is false; Then there exist two elements $b, c \in \Gamma(Ca) \cup k(Ca)$ with $b \notin \mathrm{acl}(C)$ and $c \notin \mathrm{acl}(Cb)$. Find $b' \models \mathrm{tp}(b/C)$ with $b' \notin M$. Conjugating by an element of $\mathrm{Aut}(\mathcal{U}/C)$ we may assume $b \notin M$. Find $c' \models \mathrm{tp}(c/Cb)$ with $c' \notin \mathrm{acl}(Cb)$. Conjugating by an element of $\mathrm{Aut}(\mathcal{U}/Cb)$ we may assume $c \notin \mathrm{acl}(Mb)$. But over M, U is definably isomorphic to a subset of a quotient of K. So there exists an element $a' \in K$ with $\mathrm{rk}_{\mathbb{Q}}(\Gamma(Ma')/\Gamma(M)) + \mathrm{trdeg}(k(Ma')/k(M)) > 1$, a contradiction. $\square$

In [12, Section 2.3] a notion of *relative radius* of a definable subtorsor is defined. We repeat it, as it is occasionally used here. Let U be a torsor of the module A; a *subtorsor* is a subset of the form $V = u + B$, where B is a submodule of A. Note that $B = \{v - v' : v, v' \in V\}$. Let $\mathrm{Sub}(U)$ be the set of definable subtorsors of U.

We shall say that a definable 1-torsor U is *special* if it is a torsor of a 1-module A which is definably isomorphic to a quotient of R or $\mathcal{M}$ or a proper quotient of K, or equals K itself (rather than just being definably isomorphic to K). A *special unary set* is a subset of Γ of form $[0, \alpha)$ or a special 1-torsor.

LEMMA 7.12. *Let U be a C-definable special 1-torsor. Then there exists a C-definable function* rad: $\mathrm{Sub}(U) \to \Gamma$ *such that for any $V \in \mathrm{Sub}(U)$,*

$$V' \mapsto \mathrm{rad}(V')$$

is a bijective, order-preserving map between $\{V' \in \mathrm{Sub}(U): V \subseteq V'\}$ and an interval in $[0, \infty]$.

PROOF. (i) Suppose U is a 1-torsor of the 1-module A, with definable subtorsor V. This means that V, a subset of U, is a torsor of a definable submodule B of A. We put $\mathrm{rad}(V) := \mathrm{rad}(B)$, so have to define $\mathrm{rad}(B)$. Suppose first A is closed. Then for some unique γ, $B = \gamma RA$ or $\gamma \mathcal{M}A$. Then $\mathrm{rad}(V) := \gamma$. If A is open (but not definably isomorphic to a quotient of K), then a definable submodule has radius γ if it has the form γRA or $\bigcap(\delta RA: \delta > \gamma)$.

The definition of radius for a subtorsor of a torsor arising from a proper quotient of K is clear. First, any such quotient, if non-trivial, is definably isomorphic to K/R of $K/\mathcal{M}$, so we only consider these cases. If A is definably isomorphic to K/R, then a definable submodule D has radius $\mathrm{rad}(D) := \gamma$ if γ is greatest such that $\gamma RD = \{0\}$; if A is definably isomorphic to $K/\mathcal{M}$, then D has radius γ where γ is greatest such that γRD is isomorphic to $\{0\}$ or k.

Finally, if $A = K$, the definition of radius for subtorsors of U is clear. □

REMARK 7.13. We extend the definition of rad to the case when U is a $C - \infty$-definable 1-torsor which is the intersection of a chain $(U_i: i \in I)$ of definable subtorsors of some 1-torsor V, where V is definably isomorphic to a quotient of R or M. In this case, we arbitrarily fix some $i_0 \in I$, and for any definable subtorsor W of U, define $\mathrm{rad}(W)$ with respect to U_{i_0}, i.e., by regarding W as a subtorsor of the *definable* 1-torsor U_{i_0}. This device ensures that the radius lies in Γ rather than in its Dedekind completion.

If T is a closed 1-torsor, say of the closed 1-module A, we write $\mathrm{red}(T)$ for its reduction $T/\mathcal{M}A$, the quotient of T by the action of $\mathcal{M}A$. This has definably, without extra parameters, the structure of a 1-dimensional affine space over k, so is strongly minimal.

The following result of [12] shows that any element of the geometric sorts can be thought of as a sequence of realisations of unary types. If $a = (a_1, \ldots, a_m)$ is a sequence of elements from the sorts $\mathcal{G}$, we say a *is unary* if, for each $i = 1, \ldots, m$, a_i is an element of a unary set defined over $\mathrm{dcl}(a_j: j < i)$. Trivially, any finite sequence of elements of K is unary, since K itself is a unary set, as $K = K/0.R$.

PROPOSITION 7.14. [12, *Proposition 2.3.10*] *Let* $s \in \mathcal{U}$. *There is a unary sequence* (*called a* unary code) $(a_1, \ldots, a_m)$ *such that* $\mathrm{dcl}(s) = \mathrm{dcl}(a_1, \ldots, a_m)$.

The proof of Proposition 7.14 exploits the triangular form of the matrices in the identification of S_n with $B_n(K)/B_n(R)$ and T_n with $\bigcup_{i=0}^{n} B_n(K)/B_{n,m}(R)$. Take the case of S_n. The group $B_n(K)$ has a normal unipotent subgroup $U_n(K)$ (the strictly upper triangular matrices) with quotient $D_n(K)$ isomorphic to $(K^*)^n$. Now $D_n(K)/D_n(R)$ is isomorphic to Γ^n, and leads to 1-torsors of the type Γ. On the other hand $U_n(K)$ has a sequence of normal subgroups N_j with successive quotients isomorphic to the additive group of K. This leads to fibers of the form $gN_{j+1}U_n(R)/N_jU_n(R)$; these are torsors for

$$N_{j+1}U_n(R)/N_jU_n(R) \cong N_{j+1}/N_j(N_{j+1} \cap U_n(R))$$

Since $N_{j+1}/N_j \cong (K, +)$ and since $N_{j+1} \cap U_n(R) \neq (0)$, this leads to 1-torsors for modules isomorphic to K/R. Similarly T_n can be analyzed by elements of Γ and of 1-torsors for modules isomorphic to $K/\mathcal{M}$.

This shows that the statement can be improved somewhat:

PROPOSITION 7.15. *Let* $s \in \mathcal{U}$. *There is a sequence* $(a_1, \ldots, a_m)$ *and* $C_0, \ldots,$ C_{m-1} *such that* $\mathrm{dcl}(s) = \mathrm{dcl}(a_1, \ldots, a_m)$, $a_i \in C_{i-1}$, *and* C_i *is either* Γ *or* K *or an* $(a_1, \ldots, a_i)$-*definable torsor of an* R-*module isomorphic to* K/R *or* $K/\mathcal{M}$.

In particular, the unary sets involved can be chosen to be special. Alternatively, using the proof of Proposition 7.14 given in [12], we can obtain a decomposition with unaries of the form Γ, K or closed 1-torsors (belonging to *non-free* 1-generated R-modules); again, these are all special.

In [12] we developed a theory of independence for the unary types, including results on definability of types and orthogonality to the value group. Part of the goal in this monograph is to develop these results more thoroughly for n-types. We reproduce the definitions and elementary results from [12].

First, recall from Section 2 of [12] that a *Swiss cheese* is a subset of K of the form $t \setminus (t_1 \cup \cdots \cup t_m)$, where t (the *block*) is a ball of K or the whole of K, and the t_i (the *holes*) are distinct proper sub-balls of t (which could be field elements). More generally, if U is a 1-torsor, then a subset of U of the form $t \setminus (t_1 \cup \cdots \cup t_m)$ (where $t, t_1, \ldots, t_m$ are definable subtorsors) is regarded as a Swiss cheese of U, with corresponding notions of 'hole' and 'block'. If U_1, U_2 are Swiss cheeses of U, we say that they are *trivially nested* if the hole of one is equal to the block of another; in this case, $U_1 \cup U_2$ can be represented as a *single* Swiss cheese. The following lemma, partly due to Holly [14], is a consequence of quantifier elimination in ACVF.

LEMMA 7.16. [12, *2.1.2 and 2.3.3*] *Let* U *be a* C-1-*torsor*.

(i) *Let* X *be a definable subset of* U. *Then* X *is uniquely expressible as the union of a finite set* $\{A_1, \ldots, A_m\}$ *of Swiss cheeses, no two trivially nested*.

(ii) *Suppose in addition that* $C = \mathrm{acl}(C)$. *If* $a, b \in U$ *and neither of* a, b *lie in a* C-*definable proper subtorsor of* U, *then* $a \equiv_C b$.

The proof uses the fact that any definable subset of K is a Boolean combination of balls, and the observation that any definable unary set is in definable bijection with K, an interval of Γ, or a ball.

DEFINITION 7.17. Let U be an acl(C)-unary set and $a \in U$. Then a is *generic in U over C* if a lies in no acl(C)-unary proper subset of U.

In the motivating case when $a \in K$, U will be a C-definable ball (possibly K itself) or the intersection of a chain of C-definable balls. Then a is generic in U over C if and only if there is no sub-ball of U which is algebraic over C and contains a.

REMARK 7.18. (i) By Lemma 7.16, if a, b are generic over C in a C-unary set U, then $a \equiv_{\mathrm{acl}(C)} b$. Thus, we may talk of *the generic type of U* (over C) as the type of an element of U which is generic over C. This uniqueness, like Lemma 7.19, follows from Lemma 7.16.

(ii) If T is a closed 1-torsor then the above notion of genericity for the strongly minimal 1-torsor red(T) agrees with that from stability theory. That is, if T is C-definable then $t \in \mathrm{red}(T)$ is generic over C if it does not lie in any C-definable finite subset of T. Also, suppose T is a C-definable closed 1-torsor, and a is generic in red(T) over C. Then all elements of a have the same type over C; for otherwise, some C-definable subset of T intersects infinitely many elements of red(T) in a proper non-empty subset, contradicting Lemma 7.16 (i).

(iii) We adapt slightly the above language, by saying that if $\gamma_0 \in \Gamma(C)$, then γ is *generic over C below γ_0* if for any $\varepsilon \in \Gamma(C)$, if $\varepsilon < \gamma_0$ then $\varepsilon < \gamma$. That is, γ is generic in the unary set $[0, \gamma_0)$.

LEMMA 7.19. [12, *Lemma* 2.3.6] *Suppose* $C = \mathrm{acl}(C)$, *and a is an element of a C-unary set U. Then a realises the generic type over C of a unique C-unary subset V of U.*

To see this in the case when U is a 1-torsor, let V be the intersection of the set of C-definable subtorsors of U containing a.

It follows from the lemma that if $C = \mathrm{acl}(C)$ and a is a field element, then $\mathrm{tp}(a/C)$ is the generic type over C of a unary set. Likewise if $\gamma \in \Gamma(C)$ and a is a ball (say closed) of radius γ, then a lies in the C-unary set $K/\gamma R$, so realises the generic type over C of some unary set; namely, the intersection of the C-definable subtorsors of $K/\gamma R$ which contain a.

LEMMA 7.20. [12, *Lemma* 2.3.8] *Let C be any set of parameters.*

(i) *If p is the generic type over C of a C-definable unary set, then p is definable over C.*

(ii) *Let $\{U_i : i \in I\}$ be a descending sequence of C-definable subtorsors of some C-1-torsor U, with no least element, and let p be the generic type over $\mathcal{U}$ of field elements of $\bigcap(U_i : i \in I)$. Then p is not definable.*

For example, in (i), suppose U is the closed ball R, and $\varphi(x, y)$ is some formula. There is some n_φ such that for each c, $\varphi(x, c) \in p$ if and only if, for all but at most n_φ elements α of k, $\varphi(x, c)$ holds for all elements of R with residue α. Thus, then $(d_p x)(\varphi(x, y))$ is just

$$\exists \xi_1 \ldots \exists \xi_{n_\varphi+1} \left(\bigwedge_{i=1}^{n_\varphi+1} \xi_i \neq \xi_j \wedge (\forall x \in R)\left(\bigvee_{i=1}^{n_\varphi+1} \mathrm{res}(x) = \xi_i \to \varphi(x, y) \right) \right).$$

DEFINITION 7.21. Let a be an element of a unary set, and C, B be sets of parameters with $C = \mathrm{acl}(C) \subset \mathrm{dcl}(B)$. We say that a is *generically independent from B over C*, and write $a \underset{C}{\overset{g}{\downarrow}} B$, if either $a \in \mathrm{acl}(C)$, or, if a is generic over C in a C-unary set U, it remains generic in U over B.

Without extra assumptions, this notion is not symmetric. If $a \underset{C}{\overset{g}{\downarrow}} B$ and $b \in B$, we might not have $b \underset{C}{\overset{g}{\downarrow}} Ca$. See Example 8.4 for a counterexample. However, the following easy result gives a kind of stationarity principle for generic extensions of unary types, and yields that they have invariant extensions. It will be extended in the next chapter to arbitrary types.

PROPOSITION 7.22. [12, *Proposition 2.5.2*] *Let B, C be sets of parameters with $C = \mathrm{acl}(C) \subseteq \mathrm{dcl}(B)$, and let p be the type of an element of a C-unary set U. Then there is a unique unary type q over B extending p such that if* $\mathrm{tp}(a/B) = q$ *then* $a \underset{C}{\overset{g}{\downarrow}} B$.

7.7. One-types orthogonal to Γ

In Section 2.4 of [12] there is a complete description of definable functions from Γ to $\mathcal{U}$. We shall not need these in their general form, but we quote a result which underpins some of the orthogonality discussion below.

PROPOSITION 7.23. [12, *Proposition 2.4.4*] *Let B be a set of parameters, U a B-1-torsor, $\alpha, \gamma \in \Gamma$, and t be a subtorsor of U of radius γ (possibly 0) with $t \in \mathrm{acl}(B\alpha) \setminus \mathrm{acl}(B)$. Then $\gamma \in \mathrm{dcl}(B\alpha)$ and there is an $s \in \mathrm{acl}(B)$ (a subtorsor of U) with $\mathrm{rad}(s) < \gamma$, such that $t \in \{B_{\leq\gamma}(s), B_{<\gamma}(s)\}$.*

DEFINITION 7.24. Let $C = \mathrm{acl}(C)$, and a be an element of a C-unary set. We write $\mathrm{tp}(a/C) \perp \Gamma$, and say $\mathrm{tp}(a/C)$ is *orthogonal* to Γ if, for any algebraically closed valued field M such that $C \subseteq \mathrm{dcl}(M)$ and $a \underset{C}{\overset{g}{\downarrow}} M$, we have $\Gamma(M) = \Gamma(Ma)$.

This definition will be extended to arbitrary types in Chapter 10. To see the reason for going up to a model M, suppose $a \in R$ is not algebraic over $\emptyset$, and $s := \ulcorner B_{<1}(a) \urcorner$. Then $\Gamma(s) = \Gamma(sa) = \{0, 1\}$. However, if M is any model with $s \in \mathrm{dcl}(M)$, then M contains a field element b in the ball coded by s. Now if $a \underset{b}{\overset{g}{\downarrow}} M$ then $\gamma := |b-a| \in \Gamma(Ma) \setminus \Gamma(M)$. Indeed if $\gamma \in \mathrm{dcl}(M)$ then

$B_{\leq\gamma}(b)$ is a proper sub-ball of $B_{<1}(a)$ containing a, contradicting genericity. This argument shows that the generic type of an open ball, or the intersection of a chain of balls with no least element, cannot be orthogonal to Γ. In fact, we have the following.

LEMMA 7.25. [12, *Lemma 2.5.5*] *Let* $C = \mathrm{acl}(C)$ *and* $a \notin C$ *lie in a* C-*unary set* U. *Then the following are equivalent*:

(i) a *is generic over* C *in a closed subtorsor of* U *defined over* C.
(ii) $\mathrm{tp}(a/C) \perp \Gamma$.

Furthermore, if $A := \mathrm{acl}(Ca)$ *then condition*

(iii) $\mathrm{trdeg}(k(A)/k(C)) = 1$

implies both (i) *and* (ii). *If in addition* $C = \mathrm{acl}(C \cap K)$, *then* (i), (ii) *are equivalent to* (iii).

The following lemmas follow in [12] from the analysis of definable functions with domain Γ.

LEMMA 7.26. [12, *Lemma 3.4.12*] *If* $C = \mathrm{acl}(C)$, *and* $\alpha \in \Gamma$, *then* $\mathrm{acl}(C\alpha) = \mathrm{dcl}(C\alpha)$.

LEMMA 7.27. [12, *Lemma 2.5.6*] *If* T *is a* C-1-*torsor which is not a closed* 1-*torsor, then the following are equivalent*:

(i) *no proper subtorsor* T' *of* T *is algebraic over* C;
(ii) *for all* a *generic in* T, $\Gamma(C) = \Gamma(Ca)$.

To see the direction (i) $\Rightarrow$ (ii), suppose that a is generic in T and $\delta \in \Gamma(Ca)\backslash\Gamma(C)$. Then there is a C-definable function $T \to \Gamma$ with $f(a) = \delta$, and $f^{-1}(\delta)$ is a proper $C\delta$-definable subset of T. It follows easily from Lemma 7.16 that there is a $C\delta$-definable proper subtorsor T_δ of T. By Proposition 7.23, T_δ is a neighbourhood of a subtorsor T' of T which is definable over C.

DEFINITION 7.28. [12, *Definition 2.5.9*] If $C = \mathrm{acl}(C)$ and a lies in some unary set, we say that $\mathrm{tp}(a/C)$ is *order-like* if a is generic over C in a C-unary set which is either (i) contained in Γ, or (ii) an open 1-torsor, or (iii) the intersection of a chain of C-definable 1-torsors with no least element, such that this intersection contains some proper C-definable subtorsor.

LEMMA 7.29. [12, *Remark 2.5.10*] (i) *Suppose that* $\mathrm{tp}(a/C)$ *is order-like* (*and in case* (ii) *of the last definition, assume also* $C = \mathrm{acl}(C \cap K)$). *Then* $\Gamma(C) \neq \Gamma(Ca)$.

(ii) *Suppose that* a *lies in some unary set but* $\mathrm{tp}(a/C)$ *is not order-like. Then* $\Gamma(C) = \Gamma(Ca)$.

PROOF. (i) This follows from Lemma 7.27 (ii) $\Rightarrow$ (i).

(ii) Apply Lemma 7.25 if $\mathrm{tp}(a/C)$ is the generic type of a closed ball, and Lemma 7.27 otherwise.   □

The next lemma gives a symmetry property of $\downarrow^g$ to be extended in Propositions 8.21 and 8.22 below. A special easy case (which plays a role in the proof) is when a and b are each generic over C in a C-definable closed ball.

LEMMA 7.30. [12, *Lemma 2.5.11*] *Suppose* $C = \mathrm{acl}(C)$ *and* a *and* b *are respectively elements of the* C-1-*torsors* U *and* V. *Assume that at least one of* $\mathrm{tp}(a/C)$, $\mathrm{tp}(b/C)$ *is not order-like. Then* $a \downarrow^g_C \mathrm{acl}(Cb)$ *if and only if* $b \downarrow^g_C \mathrm{acl}(Ca)$.

7.8. Generic bases of lattices

We shall need repeatedly a notion of *generic basis* for a lattice, introduced in Section 3.1 of [12]. For $s \in S_n$, $B(s) := \{a \in (K^n)^n : a = (a_1,\ldots,a_n), \Lambda(s) = Ra_1 + \cdots + Ra_n\}$, the set of all bases of $\Lambda(s)$. We shall describe an $\mathrm{Aut}(\mathcal{U}/C)$-invariant extension q_s of the partial type $B(s)$ over C, where $s \in \mathrm{dcl}(C)$. As $\mathrm{red}(s)^n$ is a definable set of Morley rank n^2 and degree 1 in the structure $\mathrm{VS}_{k,C}$, it has a unique generic type (in the sense of stability theory) $q_{\mathrm{red}(s)^n}$ over $\mathcal{U}$. Now $a = (a_1,\ldots,a_n) \models q_s$ if and only if $(\mathrm{red}(a_1),\ldots,\mathrm{red}(a_n)) \models q_{\mathrm{red}(s)^n}$. To show that q_s is complete, observe that there is a $\mathcal{U}$-definable isomorphism $\Lambda(s) \to R^n$. Thus we may suppose $\Lambda(s) = R^n$. Now $q_{\mathrm{red}(s)^n}$ is just the type over $\mathcal{U}$ of a generic element $(\beta_1,\ldots,\beta_{n^2})$ of k^{n^2}. It follows easily from Remark 7.18 (ii) that for such a sequence, any two tuples $(b_1,\ldots,b_{n^2})$, where $\mathrm{res}(b_i) = \beta_i$ for each i, have the same type. This gives completeness of q_s, and, along with the invariance of $q_{\mathrm{red}(s)^n}$, yields invariance of q_s.

We call a realisation of $q_s|C$ a *generic basis* of $\Lambda(s)$ over C or *generic resolution* of s over C, and also talk of a generic resolution over C of a sequence $s_1,\ldots,s_m$ of codes of C-definable lattices; the latter is a sequence $b_1,\ldots,b_m$, where each b_i is a generic basis of $\Lambda(s_i)$ over $Cb_1\ldots b_{i-1}$. The order of the sequence is irrelevant to genericity.

LEMMA 7.31. [12, *Remark 3.1.1*] *Let* $s \in S_n \cap \mathrm{dcl}(C)$, *let* $B \supseteq C$, *and suppose that* $a \models q_s|_B$. *Then* $\Gamma(B) = \Gamma(Ba)$.

We omit the proof, but for example, if B is a model then $\Lambda(s)$ is interdefinable with R^n and a realisation of $q_s|B$. The latter is just a generic sequence of realisations of the closed ball R, so by Lemma 7.25 does not extend the value group .

CHAPTER 8

SEQUENTIAL INDEPENDENCE

In this chapter, we extend Definition 7.21 of generic independence for an element of a unary type to a definition of sequential independence for arbitrary tuples. This yields invariant extensions of arbitrary types over algebraically closed sets. For stably dominated types, we observe as a consequence that sequential independence coincides with independence defined in Part 1.

Sequential independence can equally well be defined from generic independence for one-types in an o-minimal or weakly o-minimal theory: see Example 13.3. However as discussed in the introduction, the need to work over algebraically closed sets makes it impossible here to reduce to ambient dimension one: one cannot stay within the family of algebraically closed sets while adding one point at a time. It is necessary to add the whole algebraic closure along with the new point; the reduction is only to pro-finite covers of unary sets, hence by compactness to finite covers of unary sets. It is for the same reason that pseudo-finite fields do not admit quantifier-elimination, but require quantifiers over algebraically bounded sets.

As Examples 8.4 and 8.5 show, sequential independence is not preserved under permutations of the variables, even in the field sort; the same issues would arise in the o-minimal case. However, we do obtain uniqueness of sequentially independent extensions, and in particular, this serves to show the existence of invariant extensions of *any* type over an algebraically closed set (Corollary 8.16 below). Sequential independence will be used in Chapter 10 in the definition of orthogonality to Γ. The latter turns out to be the same as stable domination, and thus leads to a more symmetric form of independence. Using sequential independence we also show, at the end of this chapter, that over an algebraically closed set C the collection of n-types which extend to a C-definable type over $\mathcal{U}$ is dense in $S_n(C)$.

For more examples involving sequential independence, see Chapter 13.1 and the end of Chapter 15.

DEFINITION 8.1. If $A \subseteq \operatorname{acl}(Ca)$ for some finite tuple a, we say that A is *finitely* acl-*generated over* C, and call a an acl-*generating sequence*. Likewise, if $A \subseteq \operatorname{dcl}(Ca)$ then A is *finitely* dcl-*generated over* C, with dcl-*generating sequence* a.

DEFINITION 8.2. Let A, B, C be sets. For $a = (a_1, \ldots, a_n)$, and $U = (U_1, \ldots, U_n)$, define $a \underset{C}{\overset{g}{\bigcup}} B$ *via* U ('a *is sequentially independent from* B *over* C *via* U') to hold if for each $i \leq n$, U_i is an acl$(Ca_1 \ldots a_{i-1}) - \infty$-definable unary set, and a_i is a generic element of U_i over acl$(BCa_1 \ldots a_{i-1})$. We allow here the degenerate case when $a_i \in \text{acl}(Ca_1 \ldots a_{i-1})$, formally putting $U_i = \{a_i\}$.

We shall say $A \underset{C}{\overset{g}{\bigcup}} B$ *via* a, U if a is an acl-generating sequence for A over C and $a \underset{C}{\overset{g}{\bigcup}} B$ *via* U (and we sometimes omit the reference to U). We say $A \underset{C}{\overset{g}{\bigcup}} B$ if $A \underset{C}{\overset{g}{\bigcup}} B$ *via* some a, U. Finally, we say $A \underset{C}{\overset{g}{\bigcup}} B$ *via any generating sequence* if for any unary acl-generating sequence a for A over C, with a chosen from A, we have $a \underset{C}{\overset{g}{\bigcup}} B$ *via* U for some U.

REMARK 8.3. Definition 8.2 also makes sense for transfinite tuples $a = (a_i : i < \lambda)$. If $U = (U_i : i < \lambda)$, we say that $a \underset{C}{\overset{g}{\bigcup}} B$ if for each $i < \lambda$, U_i is an acl$(Ca_j : j < i) - \infty$-definable unary set, and a_i is generic in it over acl$(BCa_j : j < i)$. We shall refer to such a as a *unary transfinite sequence*. For concreteness, we generally work with finite sequences, and make occasional comments indicating how most of the results lift to the transfinite version. The infinite version is used in the proof of Theorem 15.5.

The following cautionary examples show that $\underset{}{\overset{g}{\bigcup}}$ is not symmetric in general, and that the relation $a \underset{C}{\overset{g}{\bigcup}} B$ can depend on the order of $a = (a_1 \ldots, a_n)$.

EXAMPLE 8.4. (Failure of symmetry) Suppose that b is generic over $\emptyset$ in the open ball $\mathcal{M} = B_{<1}(0)$ and a is generic over b, also in $B_{<1}(0)$. Then $a \underset{\emptyset}{\overset{g}{\bigcup}} b$. But $|b| < |a|$, so $b \underset{\emptyset}{\overset{g}{\not\bigcup}} a$, as the a-definable ball $B_{\leq|a|}(0)$, which is smaller than $B_{<1}(0)$, contains b.

EXAMPLE 8.5. (Dependence on the enumeration) Let $b_1 \in \mathcal{M}$ and $b_2 \in 1 + \mathcal{M}$. Choose $a_1 \in \mathcal{M}$ generic over $b_1 b_2$, and a_2 generic in $1 + \mathcal{M}$ over $a_1 b_1 b_2$. Then $a_1 a_2 \underset{C}{\overset{g}{\bigcup}} b_1 b_2$. However, by genericity of a_2, $|a_1 - b_1| < |a_2 - b_2| < 1$, so $a_1 \underset{Ca_2}{\overset{g}{\not\bigcup}} b_1 b_2$ as $a_1 \in B_{\leq|a_2-b_2|}(0)$, and hence $a_2 a_1 \underset{C}{\overset{g}{\not\bigcup}} b_1 b_2$.

For more complicated examples of sequential independence, see Chapter 13.1.

Despite these examples, sequential independence behaves in some ways as one would expect of an independence relation, and has useful properties: there is transitivity and monotonicity for the parameters on the right (by the next lemma), and sequentially independent extensions always exist (by Proposition 8.8 (i)).

LEMMA 8.6. (i) *If* $C \subseteq B \subseteq D$, *then* $A \underset{C}{\overset{g}{\bigcup}} D$ *via* a, U *if and only if* $A \underset{C}{\overset{g}{\bigcup}} B$ *via* a, U *and* $A \underset{B}{\overset{g}{\bigcup}} D$ *via* a, U.

(ii) *If a_1, U_1 are k-tuples and a_2, U_2 are $(n - k)$-tuples, and $a = a_1 a_2$, $U = U_1 U_2$, then $a \mathop{\underset{C}{\overset{g}{\smile}}} B$ via U if and only if $a_1 \mathop{\underset{C}{\overset{g}{\smile}}} B$ via U_1 and $a_2 \mathop{\underset{Ca_1}{\overset{g}{\smile}}} B$ via U_2.*

PROOF. This is immediate from the definition.   □

LEMMA 8.7. *For any A, B with $B \subseteq A$ and A finitely acl-generated over B, we have $A \mathop{\underset{B}{\overset{g}{\smile}}} B$.*

PROOF. By transitivity, it suffices to do this when $A = \mathrm{acl}(Ba)$ for some single $a \in \mathcal{U}$, and this is immediate except when a is a code for an element L of $\mathcal{S} \cup \mathcal{T}$. However, by Proposition 7.14, in this case L has a unary code $(e_1, \ldots, e_n)$ with each e_i in a unary set V_i definable over $\mathrm{acl}(e_j : j < i)$. If V_i is a 1-torsor, choose U_i to be the intersection of the $\mathrm{acl}(Be_j : j < i)$-definable subtorsors of V_i which contain e_i, and argue similarly if V_i is an interval of Γ. Then U_i is an $\mathrm{acl}(Be_j : j < i)$-$\infty$-definable unary set, and e_i is generic in U_i over these parameters.   □

PROPOSITION 8.8. *Let A, B, C be sets, and $a = (a_1, \ldots, a_n)$ be in A such that $A \subseteq \mathrm{acl}(Ca_1 \ldots a_n)$, and each a_i lies in a unary set defined over $\mathrm{acl}(Ca_j : j < i)$.*

(i) *There are A', a' such that $A'a' \equiv_C Aa$ and $A' \mathop{\underset{C}{\overset{g}{\smile}}} B$ via a'.*

(ii) *There is $B' \equiv_C B$ such that $A \mathop{\underset{C}{\overset{g}{\smile}}} B'$ via a.*

PROOF. (i) This is by induction on i. The case $i = 1$ follows from Proposition 7.22. For any $1 \leq j \leq n$ write $A_j = \mathrm{acl}(Ca_1 \ldots a_j)$, $A'_j = \mathrm{acl}(Ca'_1 \ldots a'_j)$. Suppose for the inductive hypothesis that for some i we have $A'_i \mathop{\underset{C}{\overset{g}{\smile}}} B$ via $(a'_1, \ldots, a'_i)$ and $\sigma_i \in \mathrm{Aut}(\mathcal{U}/C)$ with $\sigma_i(A_i) = A'_i$, $\sigma_i(a_j) = a'_j$ for $j = 1, \ldots, i$. By Proposition 7.22, there is a'_{i+1} such that $A_i a_{i+1} \equiv_C A'_i a'_{i+1}$ and $a'_{i+1} \mathop{\underset{A'_i}{\overset{g}{\smile}}} B$. Now σ_i extends to $\sigma'_i : A_i \cup \{a_{i+1}\} \to A'_i \cup \{a'_{i+1}\}$, which extends to $\sigma_{i+1} : A_{i+1} \to A'_{i+1}$. Finally, put $A' := A'_n$.

(ii) First find $A'a'$ as in (i). Let $\tau \in \mathrm{Aut}(\mathcal{U}/C)$ be such that $\tau(a) = a'$ (so $\tau(A) = A'$). Then as $A' \mathop{\underset{C}{\overset{g}{\smile}}} B$ via a', we have $A \mathop{\underset{C}{\overset{g}{\smile}}} \tau^{-1}(B)$ via a. Now put $B' := \tau^{-1}(B)$.   □

Note that the transfinite analogues of 8.6, 8.7 and 8.8 also hold.

We now use sequential independence to show that all types over C have $\mathrm{Aut}(\mathcal{U}/C)$-invariant extensions. The essential point is that the uniqueness or 'stationary' statement of Proposition 7.22 also extends to sequentially independent extensions of n-types, but requires more work because of problems with algebraic closure. First, we need three lemmas. The first is immediate, the second is a restatement of Proposition 3.4.13 of [12], and the third is the key to the uniqueness result.

LEMMA 8.9. *Let p be a type over $\mathcal{U}$, and let $g_1, \ldots, g_r$ be definable functions which agree on p. Then there is a $\ulcorner\{g_1, \ldots, g_r\}\urcorner$-definable function g which agrees with the g_i on p.*

PROOF. Define g by putting $g(x) = y$ if $g_1(x) = \cdots = g_r(x) = y$, and $g(x) = 0$ (or some $\emptyset$-definable element of the appropriate sort) otherwise. $\square$

The following lemma is taken from [12]. Recalling that generic types of closed unary sets are stably dominated, an independent proof of the closed case can be found in Chapter 6. The other cases reduce to the closed case using the classification in [12] of definable maps from Γ.

LEMMA 8.10. [12, *Proposition* 3.4.13] *Let* $\mathrm{acl}(C) = C$, *and let* U *be a* C-*unary set. Let* f *be a definable function* (*not necessarily* C-*definable*) *with range in* U *such that for all* $x \in U$ *we have* $f(x) \in \mathrm{acl}(Cx)$. *Then there is a* C-*definable function* h *with the same germ on* U *as* f.

LEMMA 8.11. *Suppose* $C \subseteq B$, *with* $\mathrm{acl}(C) = \mathrm{dcl}(C)$, *and* $a := (a_1, \ldots, a_n)$, *with* $a \underset{C}{\overset{g}{\smile}} B$. *Suppose also* $s \in U$ *with* $s \in \mathrm{dcl}(CaB) \cap \mathrm{acl}(Ca)$. *Then* $s \in \mathrm{dcl}(Ca)$.

PROOF. We shall prove the lemma by induction (over all a, s, B, C) on the least m such that there is a sequence $d = (d_1, \ldots, d_m)$ such that $d \underset{C}{\overset{g}{\smile}} B$, $d \in \mathrm{dcl}(Ca)$, and $a \in \mathrm{acl}(Cd)$. Clearly the result holds if $m = 0$.

So suppose we have $d = (d_1, \ldots, d_m)$ satisfying the above with respect to a. We first show that a may be replaced by d. Let F_0 be the finite set of conjugates of a over CBd. Since $s \in \mathrm{dcl}(CBa)$, there is a CBd-definable function h on F_0 with $h(a) = s$. The graph of h is the finite set of conjugates of (a, s) over CBd, and these are also conjugate over the smaller set Cd. Hence, as (a, s) is algebraic over Cd, all its conjugates over CBd are algebraic over Cd, so $\ulcorner h \urcorner \in \mathrm{acl}(Cd)$. Let $(t_1, \ldots, t_r)$ be a code in U for $\ulcorner h \urcorner$ (which exists by elimination of imaginaries). Now (d, t_i) satisfies the hypotheses in the statement of the lemma for (a, s). If we could prove that $t_i \in \mathrm{dcl}(Cd)$ for each i, then $\ulcorner h \urcorner \in \mathrm{dcl}(Cd)$, and hence $s = h(a) \in \mathrm{dcl}(Cad) = \mathrm{dcl}(Ca)$, as required. Thus, in the assumptions we now replace a by d, and assume $s \in \mathrm{dcl}(CBd) \cap \mathrm{acl}(Cd)$, and must show $s \in \mathrm{dcl}(Cd)$.

As $s \in \mathrm{dcl}(CBd)$, there is a $CBd_1 \ldots d_{m-1}$-definable partial function g with $g(d_m) = s$. Put $C' := \mathrm{acl}(Cd_1 \ldots d_{m-1})$. By minimality of m, $d_m \notin C'$. Let $p := \mathrm{tp}(d_m/C')$, so p is the generic type of a unary set U which is ∞-definable over C'. We shall assume U is not a unary subset of Γ, for otherwise we could adjust the argument below, replacing g by a function $g^* \colon K \to U$, where $g^*(x) = g(|x|)$. Thus, we may suppose that U is the intersection of a chain $(U_i : i \in I)$ of C'-definable 1-torsors which are all subtorsors of U_0, say; possibly $U_i = U_0$ for each i. Let $A' := C' \cap \mathrm{dcl}(Cd_1 \ldots d_m)$. For each i, U_i has just finitely many conjugates over $Cd_1 \ldots d_{m-1}$, so if U_i' is such a conjugate, we cannot have one of U_i, U_i' containing the other. Given two subtorsors of U_0, either one contains the other or they are incomparable. Hence, as $d_m \in U_i$ for each i, it follows that each U_i is A'-definable. That is, p is the generic type of a 1-torsor (namely U) which is ∞-defined over A'.

By Lemma 8.10 there is a C'-definable function g' with the same p-germ as g, so $g'(d_m) = s$. Let $g_1 = g', \ldots, g_r$ be the conjugates of g' over BA'. As g is defined over $BCd_1 \ldots d_{m-1} \subseteq BA'$, and g and g_1 have the same p-germ, also g and g_i have the same p-germ for all i. By Lemma 8.9 that there is a $\ulcorner \{g_1, \ldots, g_r\} \urcorner$-definable function h which has the same p-germ as all the g_i. As $\{g_1, \ldots, g_r\}$ is a set of conjugates over BA', $\ulcorner h \urcorner \in \mathrm{dcl}(BA')$. Also, as $\ulcorner g_1 \urcorner \in \mathrm{dcl}(C')$ and $g_i \equiv_{Cd_1 \ldots d_{m-1}} g_1$, $\ulcorner g_i \urcorner \in \mathrm{dcl}(C')$ for each i, so $\{g_1, \ldots, g_r\}$ is C'-definable, and hence $\ulcorner h \urcorner \in C' = \mathrm{acl}(Cd_1 \ldots d_{m-1})$. Choose finite unary $e \in A'$ with $(d_1, \ldots, d_{m-1})$ as an initial segment so that $\ulcorner h \urcorner \in \mathrm{dcl}(CBe)$; then $e \underset{C}{\overset{g}{\smile}} B$, as $e \in \mathrm{acl}(Cd_1 \ldots d_{m-1})$. Let $(t_1, \ldots, t_\ell)$ be a code in $\mathcal{U}$ for $\ulcorner h \urcorner$. Then, by the induction hypothesis (with e replacing a, the t_i replacing s, and $(d_1, \ldots, d_{m-1})$ replacing d), we have $t_i \in \mathrm{dcl}(Ce)$ for each i, so $\ulcorner h \urcorner \in \mathrm{dcl}(Ce)$. Since $Ce \subseteq \mathrm{dcl}(Cd_1 \ldots d_m)$, and $h(d_m) = g(d_m)$, $s = h(d_m) \in \mathrm{dcl}(Cd_1 \ldots d_m)$, as required. $\qquad\square$

THEOREM 8.12. *Suppose $C \subseteq B$ with $C = \mathrm{acl}(C)$, and $a = (a_1, \ldots, a_n)$, $a' = (a'_1, \ldots, a'_n)$ with $a \equiv_C a'$, $a \underset{C}{\overset{g}{\smile}} B$ and $a' \underset{C}{\overset{g}{\smile}} B$. Then $a \equiv_B a'$.*

PROOF. This is by induction on n, and the case $n = 1$ comes from Proposition 7.22. We assume the result holds for $i - 1$, so by applying an automorphism we may assume $a_1 = a'_1, \ldots, a_{i-1} = a'_{i-1}$. Then a_i is generic in a unary set $U = \bigcap(U_j : j \in J)$ defined over $\mathrm{acl}(Ca_1 \ldots a_{i-1})$. Here the U_j are $\mathrm{acl}(Ca_1 \ldots a_{i-1})$-definable unary sets; we allow the case J infinite as well as $|J| = 1$, and in the latter case $U = U_1$ is allowed to be a point. Let $U' = \bigcap(U'_j : j \in J)$ be the conjugate of U over $\mathrm{acl}(Ca_1 \ldots a_{i-1})$ containing a'_i. For each $j \in J$, let c_j be a code in $\mathcal{U}$ for the set of conjugates of U_j over $Ca_1 \ldots a_{i-1}B$. Then $c_j \in \mathrm{dcl}(Ca_1 \ldots a_{i-1}B) \cap \mathrm{acl}(Ca_1 \ldots a_{i-1})$, so by Lemma 8.11, $c_j \in \mathrm{dcl}(Ca_1 \ldots a_{i-1})$. It follows that since U'_j is a conjugate of U_j over $Ca_1 \ldots a_{i-1}$, then $U_j \equiv_{Ba_1 \ldots a_{i-1}} U'_j$. Hence by compactness and saturation, applying an automorphism over $Ba_1 \ldots a_{i-1}$, we may suppose $U = U'$. Now by Proposition 7.22 we have $a_i \equiv_{Ba_1 \ldots a_{i-1}} a'_i$, as required. $\qquad\square$

We list a number of corollaries. We shall refer to them collectively as 'uniqueness' results (for sequentially independent extensions). In particular, we obtain the existence of invariant extensions of types over algebraically closed sets.

COROLLARY 8.13. *Suppose $C = \mathrm{acl}(C)$, and A, A', B, B' all contain C with $A \underset{C}{\overset{g}{\smile}} B$ via a, $A' \underset{C}{\overset{g}{\smile}} B'$ via a', $Aa \equiv_C A'a'$ and $B \equiv_C B'$. Then $AaB \equiv_C A'a'B'$.*

PROOF. Using an automorphism over C, we may suppose that $B = B'$. Let a_1, a'_1 be any finite sequences from A, A' respectively such that $aa_1 \equiv_C a'a'_1$. Then $aa_1 \underset{C}{\overset{g}{\smile}} B$ and $a'a'_1 \underset{C}{\overset{g}{\smile}} B$. Thus, by Theorem 8.12, $aa_1 \equiv_B a'a'_1$. $\qquad\square$

COROLLARY 8.14. *The analogue of Corollary 8.13 also holds when $a = (a_i : i < \lambda)$ is a transfinite sequence.*

PROOF. Again, we may suppose $B = B'$. For each $i < \lambda$, let $A(i) = \mathrm{acl}(Ca_j : j < i)$. As usual, we view the $A(i)$ as sequences. We prove inductively that $A(i)a_i \equiv_B A(i)'a_i'$.

Assume that this holds for all $j < i$. If i is a limit, it follows that $A(i) \equiv_B A(i)'$. Thus, we may suppose $A(i) = A(i)'$. Then as $a_i \equiv_{A(i)} a_i'$ and $A(i)$ is algebraically closed, $A(i)a_i \equiv_B A(i)a_i'$ by Proposition 7.22.

Suppose now $i = k + 1$ is a successor ordinal. Again, we may suppose $A(k) = A(k)'$. Now $a_k a_{k+1} \equiv_{A(k)} a_k' a_{k+1}'$ and $a_k a_{k+1} \underset{A(k)}{\overset{g}{\cup}} B$ and $a_k' a_{k+1}' \underset{A(k)}{\overset{g}{\cup}} B$. Hence, by Corollary 8.13, $A(k+1)a_{k+1} \equiv_B A(k+1)'a_{k+1}'$, as required. $\qquad\square$

The next two corollaries also have analogues where a is indexed by an infinite ordinal.

COROLLARY 8.15. *Let $C = \mathrm{acl}(C)$, let M be a model containing C, and suppose $A \underset{C}{\overset{g}{\cup}} M$ via $(a_0, \ldots, a_{n-1})$. Then any automorphism of M over C is elementary over A.*

PROOF. Let g be an automorphism of M over C, and let $(a_\alpha : \alpha < \lambda)$ be an enumeration of A, with $A \underset{C}{\overset{g}{\cup}} M$ via $(a_0, \ldots, a_{n-1})$. There is a sequence $(a_\alpha' : \alpha < \lambda)$ such that g extends to an elementary map $a_\alpha \mapsto a_\alpha'$. Now $(a_\alpha : \alpha < \lambda) \equiv_C (a_\alpha' : \alpha < \lambda)$, and $(a_\alpha' : \alpha < \lambda) \underset{C}{\overset{g}{\cup}} M$ via $(a_0', \ldots, a_{n-1}')$. It follows by Theorem 8.12 that the map h fixing M pointwise and taking each a_α' to a_α is elementary. Now the elementary map hg fixes A pointwise and induces g on M, as required. $\qquad\square$

COROLLARY 8.16. *Let $C = \mathrm{acl}(C)$, let a be a finite sequence, and put $A = \mathrm{acl}(Ca)$. Let M be a model containing C. Then there is $A'a' \equiv_C Aa$ such that $\mathrm{tp}(A'/M)$ is invariant under $\mathrm{Aut}(M/C)$.*

PROOF. Let d be a unary code for a. By Proposition 8.8 there is $d' \equiv_C d$ with $d' \underset{C}{\overset{g}{\cup}} M$. Now apply Corollary 8.15. $\qquad\square$

Next, we obtain some initial results on St_C and stable domination in ACVF, using the above uniqueness results.

PROPOSITION 8.17. *In ACVF, assume $C \subseteq B$. Suppose $\mathrm{tp}(A/C)$ is stably dominated, and let a be a unary acl-generating sequence for A over C (possibly transfinite). Then $A \underset{C}{\overset{d}{\cup}} B$ if and only if $A \underset{C}{\overset{g}{\cup}} B$ via a.*

PROOF. By Corollary 3.31, $\mathrm{tp}(A/\mathrm{acl}(C))$ is also stably dominated. Thus, $A \underset{C}{\overset{d}{\cup}} B \Leftrightarrow A \underset{\mathrm{acl}(C)}{\overset{d}{\cup}} B$. But $A \underset{\mathrm{acl}(C)}{\overset{d}{\cup}} B$ if and only if $\mathrm{tp}(A/B)$ has an extension to an $\mathrm{Aut}(\mathcal{U}/\mathrm{acl}(C))$-invariant type over $\mathcal{U}$ (by Lemma 3.20). Also, $A \underset{C}{\overset{g}{\cup}} B$ via a if and only if $A \underset{\mathrm{acl}(C)}{\overset{g}{\cup}} B$ via a, and by Corollary 8.15 the latter implies that tp(A/B) extends to an Aut($\mathcal{U}/$ acl(C))-invariant type over $\mathcal{U}$. The result now follows from the uniqueness part of Proposition 3.13 (ii). □

LEMMA 8.18. *Let a be generic over C in a C-1-torsor U which is not closed. Then* acl(Ca)st = acl(C).

PROOF. Suppose there is $c \in$ dcl(Ca) ∩ St$_C$ with $c \notin$ acl(C). Then there is a definable function $f : U \to$ St$_C$ with $f(a) = c$. There is also an acl(C)-definable (i.e., not just ∞-definable) 1-torsor U_0 with U as a subtorsor. Choose parameters b so that U_0 is b-definably isomorphic with a 1-torsor of the form $A/\gamma B$ (for $A \in \{K, R, \mathcal{M}\}$, $B \in \{R, \mathcal{M}\}$, and $\gamma \in \Gamma$ with $0 \le \gamma \le 1$); so, working over b, there is a natural meaning for $|x - y|$, where $x, y \in U_0$. Also pick $a' \in U$. For each $\gamma <$ rad(U), let $\hat{f}(\gamma) := \ulcorner\{f(x) : |x - a'| = \gamma\}\urcorner$. Then $\hat{f}$ is a definable function from a totally ordered set to a stable stably embedded set, so we may suppose that $\hat{f}$ is constant. Thus there is a set V such that for all $\gamma \in$ dom($\hat{f}$), $\{f(x) : x \in U \wedge |x - a'| = \gamma\} = V$. As $c \notin$ acl(C), V is infinite. Now pick $y \in V$. Then $f^{-1}(y)$ contains a proper non-empty subset of $\{x : x \in U \wedge |x - a'| = \gamma\}$ for each $\gamma \in \Gamma$. Thus $f^{-1}(y)$ is a definable subset of U which is not a finite union of Swiss cheeses, which is a contradiction. □

PROPOSITION 8.19. *In ACVF, suppose $C \subseteq B$. Suppose $a \underset{C}{\overset{g}{\downarrow}} B$ for some (possibly transfinite) tuple a, and let $A := $ acl(Ca). Then*
  (i) $A^{\text{st}} \underset{C}{\downarrow} B^{\text{st}}$.
  (ii) *$k(A)$ and $k(B)$ are linearly disjoint over $k(C)$ and $\Gamma(A)$ and $\Gamma(B)$ are $\mathbb{Q}$-linearly independent over $\Gamma(C)$.*

PROOF. By Corollary 8.16, tp(A/B) extends to an Aut($\mathcal{U}/$ acl(C))-invariant type. Parts (i) and the first assertion of (ii) follow immediately from Remark 3.14 (ii). For (ii), note that tp($\Gamma(A)/\Gamma(B)$) has an Aut($\Gamma(\mathcal{U})/\Gamma(C)$)-invariant extension over $\Gamma(\mathcal{U})$, and this implies $\Gamma(A) \cap \Gamma(B) = \Gamma(C)$. □

The following lemma, related to Lemma 8.18 will be used in the proof of Theorem 12.18.

LEMMA 8.20. *Suppose* tp(a/C) *is stably dominated. Then* tp(a/C) $\vdash$ tp($a/C\Gamma(\mathcal{U})$).

PROOF. This is immediate from the definition of stable domination, as dcl($\Gamma(\mathcal{U}) \cup C$) ∩ St$_C$ = dcl(C). □

For subsets of the field sort, the results on generic (i.e., sequential) independence from [12] for one variable and the above uniqueness results have two useful applications for extensions of transcendence degree one, or more generally for extensions increasing Γ by rank at most one. Here, rk$_\mathbb{Q}$ refers to dimension as a vector space over $\mathbb{Q}$.

Proposition 8.21. *Let $C \leq A, B$ be algebraically closed valued fields, and suppose* $\mathrm{rk}_{\mathbb{Q}}(\Gamma(AB)/\Gamma(C)) \leq 1$. *Then the condition* $A \underset{C}{\overset{g}{\downarrow}} B$ *is symmetric in A and B, and does not depend on the choice of* acl-*generating sequences.*

Proof. For convenience, in the proof we suppose that A and B have finite transcendence degree over C, but this is not essential. Let $a \in K^n$ and $b \in K^m$ with $A = \mathrm{acl}_K(Ca)$, $B = \mathrm{acl}_K(Cb)$, and $A \underset{C}{\overset{g}{\downarrow}} B$ via a. For any $1 \leq i \leq n$ and $1 \leq j \leq m$,

$$a_i \underset{\mathrm{acl}(Ca_1 \dots a_{i-1}b_1 \dots b_{j-1})}{\overset{g}{\downarrow}} b_j.$$

Since $\mathrm{rk}_{\mathbb{Q}}(\Gamma(Ca_1 \dots a_i b_1 \dots b_j)/\Gamma(Ca_1 \dots a_{i-1}b_1 \dots b_{j-1})) \leq 1$, it follows from Lemma 7.29 (i) that at least one of the types $\mathrm{tp}(a_i/\mathrm{acl}(Ca_1 \dots a_{i-1}b_1 \dots b_{j-1}))$ and $\mathrm{tp}(b_j/\mathrm{acl}(Ca_1 \dots a_{i-1}b_1 \dots b_{j-1}))$ is not *order-like*, in the sense of Definition 7.28. Hence, by Lemma 7.30, $b_j \underset{\mathrm{acl}(Ca_1 \dots a_{i-1}b_1 \dots b_{j-1})}{\overset{g}{\downarrow}} a_i$. Since this holds for each j, $b \underset{\mathrm{acl}(Ca_1 \dots a_{i-1})}{\overset{g}{\downarrow}} a_i$, and as this holds for all i, we obtain $b \underset{C}{\overset{g}{\downarrow}} a$. □

Proposition 8.22. *Let $C \leq A, B$ be algebraically closed valued fields, and suppose that* $\Gamma(C) = \Gamma(A)$, $\mathrm{trdeg}(B/C) = 1$, *and there is no embedding of B into A over C. Then*

(i) $\mathrm{tp}(A/C) \cup \mathrm{tp}(B/C) \vdash \mathrm{tp}(AB/C)$;

(ii) $\Gamma(AB) = \Gamma(B)$.

Remark 8.23. In the proposition, if $A = \mathrm{acl}_K(Ca)$ where $a \in K^n$, then by Proposition 8.8 there is some $A'a' \equiv_C Aa$ with $A' \underset{C}{\overset{g}{\downarrow}} B$ via a'. Hence (i) implies that $A \underset{C}{\overset{g}{\downarrow}} B$ via a. Likewise, if $B = \mathrm{acl}(Cb)$ then $B \underset{C}{\overset{g}{\downarrow}} A$ via b.

Proof of Proposition 8.22. (i) Let $b \in B \setminus C$. It suffices to show $b \underset{C}{\overset{g}{\downarrow}} A$. For then $B \underset{C}{\overset{g}{\downarrow}} A$ via b. Hence, if $A' \equiv_C A$ then also $B \underset{C}{\overset{g}{\downarrow}} A'$ via b. So by Corollary 8.13, $A'B \equiv_C AB$ as required.

We may suppose that b is generic over C in the $C - \infty$-definable 1-torsor U. Suppose $b \underset{C}{\overset{g}{\not\downarrow}} A$. Then there is an A-definable proper subtorsor V of U, with $b \in V$. If γ denotes the radius of V, then $\gamma \in \Gamma(A) = \Gamma(C)$. By Lemma 7.5, there is $a \in A$ with $a \in V$. As b does not embed into A over C, $b \not\equiv_C a$, so there is a C-definable proper subtorsor V' of U containing just one of a, b, and it must contain a and not b. Since $a \in V$, $V \cap V' \neq \emptyset$; hence as $b \notin V'$, $V' \subset V$. Now as V has radius $\gamma \in \Gamma(C)$ and contains V', V is C-definable (it is defined as a ball of radius γ containing V'). This contradicts that b is generic in U over C. So $b \underset{C}{\overset{g}{\downarrow}} A$.

(ii) Again, let $b \in B \setminus C$, and suppose that b is generic over C in the $C - \infty$-definable 1-torsor U. It suffices to show that if d is in the field $A(b)$, then $|d| \in \Gamma(B)$; for $\Gamma(AB)$ is the definable closure in Γ of the set of such values. We may suppose that $d = \Sigma_{i=0}^n a_i b^i = (a_1' - b) \dots (a_n' - b)$ (as A is algebraically closed). Thus, it suffices to show that $|b - a| \in \Gamma(B)$ for any

$a \in A$. If $a \notin U$, then there is a C-definable ball U' containing U but not a, and $|b - a| = |x - a|$ for any $x \in U'$, so $|b - a| \in \Gamma(A) = \Gamma(C)$. So suppose that $a \in U$. As in (i), there is a C-definable 1-torsor V containing a and excluding b. Thus, $|b - a|$ is the value of $|b - x|$ for any $x \in V$, so $|b - a| \in \Gamma(B)$, as required. $\qquad\square$

As a further corollary of the uniqueness results, we show that in the geometric sorts, Lascar strong types and strong types coincide, so the Lascar group over any parameter set is profinite (see [28] and [31] for background). The proof works in any theory in which every type has an invariant extension. This has also been noted independently by A. Ivanov [24], where there is further information on connections between invariant extensions, Lascar strong types, and Kim–Pillay strong types.

COROLLARY 8.24. *Let a, b be finite tuples, and C be a parameter set. Then a, b have the same Lascar strong type over C if and only if they have the same type over $\mathrm{acl}(C)$.*

PROOF. By elimination of imaginaries, we may suppose a, b are from $\mathcal{U}$. The direction $\Rightarrow$ is immediate, so suppose $\mathrm{tp}(a/\mathrm{acl}(C)) = \mathrm{tp}(b/\mathrm{acl}(C))$. Choose $c \equiv_{\mathrm{acl}(C)} a$ with $c \downanglefree^{g}_{C} ab$. Then choose a model $M \supset C$ with $a \downanglefree^{g}_{C} M$ and $c \downanglefree^{g}_{Ca} M$. Then $c \downanglefree^{g}_{C} Ma$ (transitivity), so $c \downanglefree^{g}_{C} M$, so by Theorem 8.12, $c \equiv_{M} a$. Thus a and c have the same Lascar strong type, and similarly b and c have the same Lascar strong type, so a and b have the same Lascar strong type (all over C). $\qquad\square$

We conclude this chapter with a result on definability of types which extends Lemma 7.20. We show that, if $C = \mathrm{acl}(C)$, the set of n-types over C which have an extension to a C-definable type over $\mathcal{U}$ is dense in the Stone space $S_n(C)$. We begin with a general lemma.

LEMMA 8.25. *Suppose that $\mathcal{U}$ is a sufficiently saturated model of some complete theory T, and $C \subset \mathcal{U}$, with $\mathrm{dcl}(C) = \mathrm{acl}(C)$. Suppose also $p = \mathrm{tp}(b/\mathcal{U})$ is C-definable, and $b' \in \mathrm{acl}(Cb)$. Then $\mathrm{tp}(bb'/\mathcal{U})$ is C-definable.*

PROOF. Clearly $\mathrm{tp}(b'/\mathcal{U}b)$ is definable. It follows that $q := \mathrm{tp}(bb'/\mathcal{U})$ is definable. Let $\varphi(yy', x)$ be a formula, and let its q-definition be $(d_q yy')\varphi(yy', x) = \theta(x, c)$, with canonical parameter $c \in \mathcal{U}^{\mathrm{eq}}$. If $\{c_i : i \in I\}$ is a complete set of conjugates of c over C, there is a corresponding set $\{q_i : i \in I\}$ of conjugates of q under $\mathrm{Aut}(\mathcal{U}/C)$. These all extend $\mathrm{tp}(b/\mathcal{U})$, so for each $i \in I$ there is b'_i such that $bb'_i \models q_i|\mathcal{U}$. As the q_i are distinct, the b'_i are also distinct. However, there are at most $\mathrm{Mult}(b'/Cb)$ such b'_i, so $|I| \leq \mathrm{Mult}(b'/Cb)$. Thus, $c \in \mathrm{acl}(C)$. $\qquad\square$

REMARK 8.26. In the last lemma we may allow b' to be a tuple of infinite length (with the same proof).

In ACVF, if $C = \mathrm{acl}(C)$, given an n-type $p(x)$ over C and a formula $\varphi(x, y)$ over C with $n = l(x)$, $m := l(y)$, write

$$S_p x \varphi(x, y) := \{b : \text{there is } a \models p \text{ such that } a \underset{C}{\overset{g}{\downarrow}} b \text{ and } \varphi(a, b)\}.$$

If the set $S_p x \varphi(x, y)$ is definable over C, let $d_p x \varphi(x, y)$ be a formula over C defining it.

THEOREM 8.27. *Let $C = \mathrm{acl}(C)$.*
(a) *Let $\theta(x)$ be an $\mathcal{L}_G$-formula over C with $l(x) = n$, and suppose $\models \exists x \theta(x)$. Suppose also that $\theta(x)$ implies that each x_i is algebraic over $C x_1 \ldots x_{i-1}$ or lies in a unary set defined over $C x_1 \ldots x_{i-1}$. Then there is an n-type p over C containing θ such that:*
  (i) *for all $\varphi(x, y)$ over C, $S_p x \varphi(x, y)$ is C-definable;*
  (ii) *p has an extension to a C-definable type p' over $\mathcal{U}$ such that for any a realising p' we have $a \underset{C}{\overset{g}{\downarrow}} \mathcal{U}$.*
(b) *The set of n-types over C which have an extension to a C-definable type over $\mathcal{U}$ is dense in the Stone space $S_n(C)$.*

PROOF. (a) (i) We use induction on $n = l(x)$. Suppose first that $n = 1$. If $\theta(x_1)$ has finitely many solutions, then choose p to be any type over C containing θ. Since p is isolated, in this case the result is immediate. So suppose $\theta(x_1)$ is non-algebraic. Then the solution set of $\theta(x_1)$ lies in a C-definable unary set U_1. We assume U_1 is a 1-torsor; the case when $U_1 \subset \Gamma$ is easier, using o-minimality of Γ. By Lemma 7.16, the solution set of $\theta(x_1)$ is a finite union of disjoint Swiss cheeses $s_1 \setminus (t_{11} \cup \cdots \cup t_{1\ell_1}), \ldots, s_k \setminus (t_{k1} \cup \cdots \cup t_{k\ell_k})$. We may assume no two of these are trivially nested, so this representation is unique. Hence, as $C = \mathrm{acl}(C)$, $\ulcorner s_i \urcorner \in C$ for each i. Let p be the generic type of s_1 over C. Then $\theta(x) \in p$. Also, p is C-definable by Lemma 7.20, and likewise the generic type of s_1 over $\mathcal{U}$ extends p and is C-definable, by the same schema. It follows that $S_p x \varphi(x, y)$ is C-definable for each φ.

Suppose now $n > 1$, and let $\theta'(x_1, \ldots, x_{n-1})$ be the formula $\exists x_n \theta(x_1, \ldots, x_n)$. By induction, there is a type p_{n-1} over C containing θ' such that for each $\varphi(x_1, \ldots, x_{n-1}, y)$, $S_{p_{n-1}}(x_1, \ldots, x_{n-1}) \varphi(x_1, \ldots, x_{n-1}, y)$ is C-definable. Let $(e_1, \ldots, e_{n-1})$ realise p_{n-1}. By the $n = 1$ case, there is $s \in \mathrm{acl}(C e_1 \ldots e_{n-1})$ and a definable 1-type q over $C e_1 \ldots e_{n-1} s$ containing $\theta(e_1, \ldots, e_{n-1}, x_n)$ so that for any formula $\varphi(x_1, \ldots, x_{n-1}, z, x_n, y)$ over C, $S_q x_n \varphi(e_1, \ldots, e_{n-1}, s, x_n, y)$ is $C e_1 \ldots e_{n-1} s$-definable, by the formula $d_q x_n \varphi(e_1, \ldots, e_{n-1}, s, x_n, y)$. Let e_n realise q, and $p = \mathrm{tp}(e_1, \ldots, e_n / C)$. Also, let $p^* = \mathrm{tp}(e_1, \ldots, e_{n-1}, s / C)$, a definable type by Lemma 8.25. Put $x' := (x_1, \ldots, x_{n-1})$. Then for any $\varphi(x', z, x_n, y)$ over C, we have $b \in S_p x \varphi(x', z, x_n, y)$

$$\iff \exists x' z \models p^* (x' z \underset{C}{\overset{g}{\downarrow}} b \wedge \exists x_n \models q \, (x_n \underset{C x'}{\overset{g}{\downarrow}} b \wedge \varphi(x', z, x_n, b)))$$

$$\Longleftrightarrow \exists x'z \models p^*(x'z \underset{C}{\overset{g}{\downarrow}} b \wedge d_q x_n \varphi(x', z, x_n, b))$$

$$\Longleftrightarrow d_{p^*} x'z (d_q x_n) \varphi(x', z, x_n, b).$$

(ii) Choose a realising p with $a \underset{C}{\overset{g}{\downarrow}} \mathcal{U}$. Put $p' := \mathrm{tp}(a/\mathcal{U})$. Then for any formula $\varphi(x_1, \ldots, x_n, y_1, \ldots, y_m)$, if $b \in \mathcal{U}^m$ then $a \underset{C}{\overset{g}{\downarrow}} b$, so $\varphi(x, b) \in p$ if and only if $b \in S_p x \varphi(x, b)$.

(b) Let $\theta(x)$ be a formula over C. By Proposition 7.14, we may assume that $\theta(x)$ asserts that x is a unary sequence. By (a)(ii), there is a C-definable type p over $\mathcal{U}$ containing $\theta(x)$. □

CHAPTER 9

GROWTH OF THE STABLE PART.

The stable part St_C of ACVF over a parameter set C grows with C. Later on we will need to study with exactitude how St_C grows when, for instance, one adds to C an element of Γ. Here we fix a tuple a and consider the stable part $\text{St}_C(a)$ of a over C, as C grows.

Recall (Definition 7.6) that in ACVF, if C is any parameter set, then $\text{VS}_{k,C}$ is the many sorted structure consisting of a sort $\text{red}(s)$ for each C-definable lattice with code s, with the induced C-definable relations. This is essentially the same as St_C, for by Proposition 7.8, every element of any C-definable, stable, stably embedded set is coded in $\text{VS}_{k,C}$. In particular, if the parameter set C is a model M of ACVF, and $s \in S_n \cap \text{dcl}(M)$, then $\text{red}(s)$ is in M-definable bijection with k^n (Lemma 7.5). Hence, over a model M, the structure $\text{VS}_{k,M}$, and hence also St_M, is essentially just k, with the M-definable relations. On the other hand when C is not a model, the essential number of sorts can be large. We shall later obtain some results by 'resolving' C, i.e., finding an appropriate a model M with $C \subset \text{dcl}(M)$, to study this situation.

We will show in this section that $\text{St}_C(a)$ is nevertheless always countably generated. First observe the following.

EXAMPLE 9.1. The residue field of a finitely generated extension of a valued field L need not be finitely generated over the residue field of L.

Indeed, let F be a field, $F((t))$ the Laurent series field with the usual valuation, trivial on F. Let $a_n \in F^{\text{alg}}$ be any sequence of elements, $s = \sum a_n t^n \in F((t))$, $L = F(s, t)$, L^h the Henselization. Then $a_0 \in L^h$; indeed if P is the minimal monic polynomial for a_0 over F, then a_0 is the unique solution of F in the neighborhood $s + \mathcal{M}$ of s. But then it follows that $\sum a_{n+1} t^n = (s - a_0)/t \in L^h$, and inductively each $a_i \in L^h$. So the residue field of L^h contains $F(a_0, a_1, \dots)$. But this is also the residue field of L.

Similarly, using the generalized power series $s' = \sum t^{n+\frac{1}{n}}$, we see that there exists a valuation on the rational function field in two variables whose value group is $\mathbb{Q}$.

Indeed, for any valued field F, any countably generated algebraic extension of the residue field of F is contained in the residue field of some finitely generated valued field extension of F. However, by Corollary 9.6 below, this is the worst that can happen.

Call a type $\mathrm{tp}(a/C)$ *stationary* if $\mathrm{dcl}(Ca) \cap \mathrm{acl}(C) = \mathrm{dcl}(C)$. Equivalently, $\mathrm{tp}(a/C) \vdash \mathrm{tp}(a/\mathrm{acl}(C))$.

LEMMA 9.2. *If* $\mathrm{tp}(a/C)$ *and* $\mathrm{tp}(b/C(a))$ *are stationary, then so is* $\mathrm{tp}(ab/C)$.

PROOF. $\mathrm{dcl}(Cab) \cap \mathrm{acl}(C) \subseteq \mathrm{dcl}(Cab) \cap \mathrm{acl}(Ca) \subseteq \mathrm{dcl}(Ca)$ by the stationarity of b/Ca; hence by the stationarity of a/C, $\mathrm{dcl}(Cab) \cap \mathrm{acl}(C) \subseteq \mathrm{dcl}(Ca) \cap \mathrm{acl}(C) = \mathrm{dcl}(C)$. $\square$

For the rest of the chapter we work in ACVF.

EXAMPLE 9.3. If $a \in \Gamma$ then $\mathrm{tp}(a/C)$ is always stationary.

PROOF. Any $\varphi \in \mathrm{tp}(a/\mathrm{acl}(C))$ is a finite union of intervals in Γ. The endpoints of these intervals, being $\mathrm{acl}(C)$-definable, are actually C-definable using the linear ordering. Hence φ is C-definable, so $\varphi \in \mathrm{tp}(a/C)$. Thus $\mathrm{tp}(a/C) \vdash \mathrm{tp}(a/\mathrm{acl}(C))$. $\square$

An extension C' of C is said to be *finite* if $C' \subseteq \mathrm{acl}(C)$ and C' is a finitely dcl-generated extension of C. In this case, $C' = \mathrm{dcl}(Ce)$ for some finite e, and the number of realizations $\mathrm{Mult}(e/C)$ of $\mathrm{tp}(e/C)$ does not depend on the choice of e. We write $\mathrm{Mult}(C'/C) = \mathrm{Mult}(e/C)$.

LEMMA 9.4. *Let* X *be a* C-*definable special unary set, and let* $a \in X$. *Then one of the following holds*:
1. *For some finite extension* C' *of* C, $\mathrm{tp}(a/C')$ *is stationary.*
2. *For some countable* $C_0 \subseteq C$, $\mathrm{tp}(a/\mathrm{acl}(C_0)) \vdash \mathrm{tp}(a/\mathrm{acl}(C))$.

PROOF. Since every type in Γ is stationary, we may therefore suppose that X is a torsor for a unary R-module M, or $X = K$. It follows (Lemma 7.12) that any proper sub-torsor Y of X can be assigned a *radius* $\mathrm{rad}(Y)$; and $Y \in \mathrm{dcl}(C, Y', \mathrm{rad}(Y))$ for any subtorsor Y' of Y.

Let W be the collection of $\mathrm{acl}(C)$-definable subtorsors of X. For $w \in W$, let $C(w) := \mathrm{dcl}(Ce)$, where e is a code in $\mathcal{U}$ for w; let $m(w) = \mathrm{Mult}(C(w)/C)$. Let $W_a = \{w \in W : a \in W\}$.

Note that if $Y' \in W$ and $Y' \subset Y$, then $Y \in \mathrm{dcl}(C, Y', \mathrm{rad}(Y))$; but $\mathrm{rad}(Y)$ is an $\mathrm{acl}(C)$-definable element of Γ, hence is C-definable; so $Y \in \mathrm{dcl}(C, Y')$, and $C(Y) \subseteq C(Y')$. Thus $m(Y) \leq m(Y')$, and if equality holds then also $C(Y) = C(Y')$.

Note also that if $w, w' \in W_a$ then so is $w \cap w'$, and $m(w \cap w') := \mathrm{Max}\{m(w), m(w')\}$. In particular, if $m(w)$ is unbounded on W_a then for all $w \in W_a$ there exists $w' \in W_a$ with $w \supset w'$ and $m(w) < m(w')$.

CASE 1. $m(w)$ is bounded on W_a.

Let Y be chosen in W_a and $m(Y)$ maximal possible. If $Y' \in W_a$ with $Y' \subset Y$, then $m(Y') \geq m(Y)$, so by maximality $m(Y') = m(Y)$, and $C(Y') = C(Y)$. This shows that any element of W_a is $C(Y)$-definable. Let

B be the intersection of all elements of W_a; then a realizes over $\mathrm{acl}(C)$ the generic type of B; it follows that $\mathrm{tp}(a/C') \vdash \mathrm{tp}(a/C)$. Thus (1) holds, with $C' := C(Y)$.

CASE 2. $m(w)$ is unbounded on W_a. Then we can find $w_1 \supset w_2 \supset \cdots \in W_a$ with $m(w_1) < m(w_2) < \ldots$. Let $B = \cap_{n=1}^{\infty} w_n$.

We claim that B contains no proper $\mathrm{acl}(C)$-definable subtorsors. Indeed, suppose B contains $w^* \in W$. Then each w_n contains w^*, and it follows as above that $w_n \in \mathrm{dcl}(C, w^*)$. So $m(w_n) \leq m(w^*)$ for each n, a contradiction.

Let C_0 be a countable substructure of C such that each w_i is $\mathrm{acl}(C_0)$-definable. Then B is ∞-definable over $\mathrm{acl}(C_0)$, and $\mathrm{tp}(a/\mathrm{acl}(C_0)) \vdash \mathrm{tp}(a/\mathrm{acl}(C))$. $\square$

COROLLARY 9.5. *Let a be any finite tuple, and C any base structure. Then there exists a countably dcl-generated algebraic extension C' of C such that $\mathrm{tp}(a/C')$ is stationary.*

PROOF. We prove this by induction on the length of a unary code for a, so may assume that $a = (a_1, \ldots, a_n)$ is unary. If $n = 1$, then either option in Lemma 9.4 clearly implies the conclusion.

So put $b = (a_1, \ldots, a_{n-1})$, and by induction let C' be a countably dcl-generated (over C) subset of $\mathrm{acl}(C)$ such that $\mathrm{tp}(b/C')$ is stationary. By Lemma 9.4 there is countably generated $D = C'(b, b_1, b_2, \ldots) \subseteq \mathrm{acl}(C'b)$ such that $\mathrm{tp}(a_n/D)$ is stationary. Put $b' := (b, b_1, b_2, \ldots)$. Now construct a tower $C_0 = C' \subseteq C_1 \subseteq C_2 \subseteq \ldots$ of finitely generated extensions of C' within $\mathrm{acl}(C')$ such that for each m, $\mathrm{tp}(b_m/bb_1 \ldots b_{m-1}, C_m) \vdash \mathrm{tp}(b_m/bb_1 \ldots b_{m-1}, \mathrm{acl}(C'))$; this is possible for as $\mathrm{tp}(b_m/C'bb_1 \ldots b_{m-1})$ is algebraic, it is isolated. Put $C'' := \bigcup_{i \geq 0} C_i$. Now $\mathrm{tp}(b'/C'') \vdash \mathrm{tp}(b'/\mathrm{acl}(C')) \vdash \mathrm{tp}(b'/\mathrm{acl}(C''))$, so $\mathrm{tp}(b'/C'')$ is stationary. Also, $\mathrm{tp}(a_n/C''b')$ is stationary, since $C'b' \subseteq C''b' \subseteq \mathrm{acl}(C'b')$ and $\mathrm{tp}(a_n/C'b')$ is stationary. Thus, by Lemma 9.2, $\mathrm{tp}(b'a_n/C'')$ is stationary, and hence so is $\mathrm{tp}(ba_n/C'')$. $\square$

COROLLARY 9.6. *Let a be any finite tuple, and C any base structure. Then $\mathrm{St}_C(a)$ is countably generated over C.*

PROOF.   CLAIM. There exists a finite $b \in \mathrm{St}_C(a)$ with $\mathrm{St}_C(a) \subseteq \mathrm{acl}(Cb)$.

PROOF OF CLAIM. Using Proposition 7.14, it suffices to show that if d lies in a unary torsor over Cb then $\mathrm{St}_C(bd) \subseteq \mathrm{acl}(\mathrm{St}_C(b), e)$ for some finite e. If $d \in \Gamma$ or if $\mathrm{tp}(d/C(b))$ is generic in some open or properly ∞-definable $C(b)$-torsor, then $\mathrm{St}_C(bd) \subseteq \mathrm{acl}(\mathrm{St}_C(b))$ by Lemma 8.18. If d is generic in a closed $C(b)$-definable torsor Y, let $\bar{d}$ be the the image of d in $\mathrm{res}(Y)$. Then d is generic in an open $C(b, \bar{d})$-torsor (the inverse image of $\bar{d}$), so $\mathrm{St}_C(bd) \subseteq \mathrm{acl}(\mathrm{St}_C(b), \bar{d})$ by the previous case; let $e = \bar{d}$.

Now let b be as in the claim, and replace C by $\mathrm{dcl}(Cb)$. In this way we may assume $\mathrm{St}_C(a) \subseteq \mathrm{acl}(C)$. On the other hand by Corollary 9.5, $\mathrm{dcl}(C(a)) \cap \mathrm{acl}(C) \subseteq \mathrm{dcl}(C, Z)$ for some countable $Z \subseteq \mathrm{acl}(C)$. So $\mathrm{St}_C(a) \subseteq \mathrm{dcl}(C, Z)$. For any tuple z from Z, $\mathrm{tp}(z/\mathrm{St}_C(a))$ is isolated, so there exists a finite $z' \in \mathrm{St}_C(a)$ such that $\mathrm{tp}(z/z') \vdash \mathrm{tp}(z/C, \mathrm{St}_C(a))$. Let $Z' = \{z' : z \in Z^m, m = 1, 2, \dots\}$. Then Z' is countable, and $\mathrm{tp}(Z/C, Z') \vdash \mathrm{tp}(Z/C, \mathrm{St}_C(a))$. So $\mathrm{St}_C(a) \subseteq \mathrm{dcl}(C, Z')$. □

PROPOSITION 9.7. *Condition (BS) holds in ACVF, so the conclusion of Proposition 6.7 holds.*

PROOF. Immediate from Corollary 9.6 and Lemma 6.5 □

CHAPTER 10

TYPES ORTHOGONAL TO Γ

In this chapter, we develop a theory of orthogonality to Γ for n-types in the sorts of $\mathcal{G}$. This extends Definition 7.24, which gives a notion of orthogonality to Γ for unary types. At first sight, Definition 10.1 appears, as with sequential independence, to be dependent on the choice of a generating sequence. We shall show that it is independent of the choice, by proving that orthogonality to Γ is the same as stable domination, in ACVF. We also extend the notion of the *resolution* of a lattice. Over a parameter set C, it is possible that a lattice is defined, but does not contain any C-definable elements. We show how to add a basis of the lattice to C, without increasing the definable closure in either the value group or the residue field. We conclude the chapter by proving that, over the value group, an indiscernible sequence is an indiscernible set (Proposition 10.16).

Some of the main results of this chapter will be superseded later on. These resolution statements, in the finitely generated case, will be majorized in Chapter 11 by a theorem yielding a canonical minimal resolution. The equivalence of orthogonality to Γ with stable domination (Theorem 10.7) is a special case of Theorem 12.18, which will be proved independently. The present hands-on proof has the merit of giving more information, when an invariant type is *not* orthogonal to Γ, of the nature of the base change needed to see the non-orthogonality. Theorem 10.16 can also be viewed in this light, since the same result over a large model will become, in Chapter 12, an immediate consequence of the theory of stable domination.

DEFINITION 10.1. If a is a unary sequence (possibly transfinite), we say that $\mathrm{tp}(a/C)$ is *orthogonal* to Γ (written $\mathrm{tp}(a/C) \perp \Gamma$) if, for any model M with $a \downarrow^g_C M$ and $C \subseteq \mathrm{dcl}(M)$, we have $\Gamma(M) = \Gamma(Ma)$.

Note that for any $C \subseteq C' \subseteq \mathrm{acl}(C)$, $\mathrm{tp}(a/C) \perp \Gamma$ if and only if $\mathrm{tp}(a/C') \perp \Gamma$. The above definition is the natural notion of orthogonality which generalises that of stability, and is based on sequentially independent extensions (for which we have existence and uniqueness). We will slightly extend this definition in 10.10, where it is given as a condition on $\mathrm{acl}(Ca)$ (a any unary sequence) and is independent of the choice of a. Recall from Lemma 7.25 that a *unary* type is orthogonal to Γ precisely if it is the generic type of a closed 1-torsor.

LEMMA 10.2. (i) *Let* $a = a_1 a_2$ *where* a_1, a_2 *are tuples, and suppose* $\operatorname{tp}(a/C) \perp \Gamma$. *Then* $\operatorname{tp}(a_1/C) \perp \Gamma$.

(ii) *Suppose* $\operatorname{tp}(a_1/C) \perp \Gamma$ *and* $\operatorname{tp}(a_2/Ca_1) \perp \Gamma$. *Then* $\operatorname{tp}(a_1 a_2/C) \perp \Gamma$.

(iii) *Suppose* $C \subseteq B$ *and* $a \underset{C}{\overset{g}{\downharpoonright}} B$. *Then* $\operatorname{tp}(a/C) \perp \Gamma$ *if and only if* $\operatorname{tp}(a/B) \perp \Gamma$.

(iv) *Suppose that for each* $i = 1, \ldots n$, a_i *is algebraic or generic in a closed unary set over* $Ca_1 \ldots a_{i-1}$. *Then* $\operatorname{tp}(a/C) \perp \Gamma$.

(v) *If* $\operatorname{tp}(a/C) \perp \Gamma$ *then* $\Gamma(C) = \Gamma(Ca)$.

PROOF. (i) Suppose $a_1 \underset{C}{\overset{g}{\downharpoonright}} M$, and choose $a_2' \equiv_{Ca_1} a_2$ with $a_2' \underset{Ca_1}{\overset{g}{\downharpoonright}} M$. Then as $a_1 a_2' \underset{C}{\overset{g}{\downharpoonright}} M$ we have $\Gamma(Ma_1a_2') = \Gamma(M)$, so $\Gamma(Ma_1) = \Gamma(M)$.

(ii) Choose M so that $a_1 a_2 \underset{C}{\overset{g}{\downharpoonright}} M$. Then $a_1 \underset{C}{\overset{g}{\downharpoonright}} M$ and $a_2 \underset{Ca_1}{\overset{g}{\downharpoonright}} M$, so $\Gamma(M) = \Gamma(Ma_1) = \Gamma(Ma_1a_2)$.

(iii) We may suppose that $C = \operatorname{acl}(C)$ and $B = \operatorname{acl}(B)$. If $\operatorname{tp}(a/C) \perp \Gamma$ and $a \underset{B}{\overset{g}{\downharpoonright}} M$ with $B \subseteq M$, then as $a \underset{C}{\overset{g}{\downharpoonright}} B$ we have $a \underset{C}{\overset{g}{\downharpoonright}} M$, so $\Gamma(M) = \Gamma(Ma)$. Thus, $\operatorname{tp}(a/B) \perp \Gamma$. Conversely, suppose $\operatorname{tp}(a/B) \perp \Gamma$ and $a \underset{C}{\overset{g}{\downharpoonright}} M$, with $C \subseteq M$. There is $M' \equiv_B M$ with $a \underset{B}{\overset{g}{\downharpoonright}} M'$, so $a \underset{C}{\overset{g}{\downharpoonright}} M'$, and by Corollary 8.13 we have $aM \equiv_C aM'$. We have $\Gamma(M'a) = \Gamma(M')$, so $\Gamma(Ma) = \Gamma(M)$.

(iv) This is immediate from Lemma 7.25 applied stepwise, together with (ii).

(v) Suppose $\gamma \notin \Gamma(C)$, with $\gamma = f(a)$, say, where f is a C-definable function. Let $\gamma' \equiv_C \gamma$. We must show $\gamma' = \gamma$. Choose a model $M \supset C$ with $a \underset{C}{\overset{g}{\downharpoonright}} M$. As $\operatorname{tp}(a/C) \perp \Gamma$, $\Gamma(M) = \Gamma(Ma)$, so $\gamma \in M$, and hence $a \underset{C}{\overset{g}{\downharpoonright}} \gamma$. Choose $a' \equiv_C a$ with $a' \underset{C}{\overset{g}{\downharpoonright}} \gamma\gamma'$. Then $a\gamma \equiv_C a'\gamma' \equiv_C a'\gamma$, so $f(a') = \gamma = \gamma'$. Hence $\gamma \in \Gamma(C)$. $\square$

Part (iv) above and Lemma 7.25 yield in particular that if $C \leq A$ are algebraically closed valued fields with $A = \operatorname{acl}_K(Ca)$, and $\operatorname{trdeg}(A/C) = \operatorname{trdeg}(k(A)/k(C))$ (informally, all of the extension is in the residue field), then $\operatorname{tp}(a/C) \perp \Gamma$. The converse is false, as shown in Example 13.1. It is also shown, in Example 13.2, that it is possible that $\operatorname{tp}(a/C) \perp \Gamma$ for all singletons $a \in \operatorname{acl}(A) \cap K$, even though $\operatorname{tp}(A/C) \not\perp \Gamma$ (in the sense of Definition 10.10 below).

Recall from Chapter 7 the notion of a *generic basis* of the lattice with code $s \in S_n \cap \operatorname{dcl}(A)$. We say that a lattice $\Lambda(s)$ defined over A is *resolved* in $B \supset A$ (or just that s is *resolved* in B) if B contains a basis for $\Lambda(s)$. If t codes an element $U(t) \in \operatorname{red}(s)$, then t is *resolved* in A if s is resolved in A and $A \cap K^n$ contains an element of $U(t)$. We say A is *resolved* if all elements of A are resolved.

DEFINITION 10.3. Let A be an $\mathcal{L}_G$-structure. Then a *generic closed resolution* of A is a structure $B = \text{acl}(A \cup \{b_i : i < \lambda\})$, where, for each $i < \lambda$, b_i is a generic basis of some lattice defined over $\text{acl}(Ab_j : j < i)$, and each lattice of B has a basis in B.

REMARK 10.4. (i) If $A = \text{acl}(A)$, then A is resolved if and only if $A = \text{dcl}(A \cap K)$. For example, if A is resolved and $t, s \in A$ with t a code for $U(t) \in \text{red}(s)$, then $U(t)$ contains some $(a_1, \dots, a_n) \in K^n$, so is defined as $U(t) = (a_1, \dots, a_n) + \mathcal{M}\Lambda(s)$.

(ii) If A is an $\mathcal{L}_G$-structure then A has a generic closed resolution $B = \text{acl}(A \cup \{b_i : i < \lambda\})$, and B may be chosen so that $\lambda \leq |A|$. Furthermore, $\Gamma(B) = \Gamma(A)$, by Lemma 7.31.

(iii) More generally, suppose $(s_1, \dots, s_m)$ is a tuple of codes for lattices in A. We definably identify $(s_1, \dots, s_m)$ with a single s which is a code for the lattice $\Lambda(s_1) \times \cdots \times \Lambda(s_m)$. Let b be a generic basis for $\Lambda(s)$ over A in the sense of Chapter 7. Then $\text{tp}(b/A) \perp \Gamma$ (treating b as a tuple of field elements). For suppose $b \underset{A}{\overset{g}{\downarrow}} M$ for some model M. By Proposition 8.19, $\text{St}_A(Ab) \underset{A}{\downarrow} \text{St}_A(M)$. It follows (from the definition of generic basis) that b is a generic basis of $\Lambda(s)$ over M, and hence, by (ii) above, $\Gamma(Mb) = \Gamma(M)$.

Let $A = \text{acl}(A)$, and suppose that all lattices of A are resolved in A. Then for any lattice $\Lambda(s)$ of A, the k-vector space $\text{red}(s)$ has a basis in A; namely, the set $\{\text{red}(b_1), \dots, \text{red}(b_n)\}$, where $\{b_1, \dots, b_n\}$ is a basis of $\Lambda(s)$. Hence, there is an A-definable bijection $\text{red}(s) \to k^n$. Let $k_{\text{rep}}(A)$ be the sub-field of $k(A)$ consisting of res(0) and elements res(a) where $a \in A$ and $|a| = 1$. Let $\mathcal{B}$ be a transcendence basis of $k(A)$ over $k_{\text{rep}}(A)$, and let $J \subset K$ with $|J| = |\mathcal{B}|$ and $\mathcal{B} = \{\text{res}(c) : c \in J\}$. Then $\text{acl}(A \cup J)$ is called a *canonical open resolution* of A.

LEMMA 10.5. *Let $A = \text{acl}(A)$, and suppose all lattices of A are resolved. Then A has a canonical open resolution B, which is unique up to A-isomorphism, and $\Gamma(A) = \Gamma(B)$. We have $B \subseteq \text{dcl}(B \cap K)$.*

PROOF. We adopt the notation above, with $B = \text{acl}(A \cup J)$. The existence part is immediate.

Let $\mathcal{B} = \{\alpha_j : j \in J\}$. Put $E := A \cap K$. Then $A = \text{acl}(E \cup \mathcal{B})$: for elements of $S_n \cap A$ lie in $\text{dcl}(E)$ by assumption, and elements of $T_n \cap A$ lie in $\text{dcl}(E \cup \mathcal{B})$ by the argument before the lemma. Also, if d_j is a field element chosen in α_j for each $j \in J$, then for each j, α_j is a generic element of k over $E \cup (\mathcal{B} \setminus \{\alpha_j\}) \cup \{d_i : i < j\}$. It follows that $\Gamma(A \cup \{d_i : i < j\}) = \Gamma(A \cup \{d_i : i \leq j\})$, so $\Gamma(A) = \Gamma(B)$.

The uniqueness assertion is also clear: the type of $(d_j : j \in J)$ over A does not depend on the choice of $\mathcal{B}$, essentially because the type of a generic sequence from R is uniquely determined. □

Before making the general connection, for invariant types, between stable domination and orthogonality to Γ, we work out an example. It will also be used in the proof.

EXAMPLE 10.6. Let $p = \operatorname{tp}(a/\mathcal{U})$ be an $\operatorname{Aut}(\mathcal{U}/C)$-invariant type of a single field element a. Suppose that $\Gamma(\mathcal{U}) = \Gamma(\mathcal{U}a)$. Then p is the generic type of a closed ball, defined over C.

PROOF. Let M be a small model, $C \subseteq M$. Now a is generic over M in the intersection V of a chain $(U_i : i \in I)$ of M-definable balls. We first show that V is closed.

Suppose first that V contains no proper M-definable ball. If $a \in B$, where B is a $\mathcal{U}$-definable ball, and B is a proper sub-ball of V, then by invariance $a \in B'$ for any $\operatorname{Aut}(\mathcal{U}/C)$-conjugate B' of B. But it is easy to find a C-conjugate B' of B with $B \cap B' = \emptyset$. This contradiction shows that there is no such ball B, so that a is generic in V over $\mathcal{U}$. This implies that V is closed; for by saturation $\mathcal{U}$ contains some field element $d \in V$, and if V is not closed then $|d - a| \in \Gamma(\mathcal{U}a) \setminus \Gamma(\mathcal{U})$, which is impossible.

In the other case, V contains a proper M-definable ball, hence a point $e \in M$. Let $\gamma = |e - a|$, and let $B_{\leq \gamma}(e) = \{x : |x - e| \leq \gamma\}$. Then $\gamma \in \Gamma(Ma) \subset \Gamma(\mathcal{U})$, and γ is fixed by $\operatorname{Aut}(\mathcal{U}/M)$, so $\gamma \in M$. Thus $B_{\leq \gamma}(e)$ is M-definable; since $a \in B_{\leq \gamma}(e)$ we must have $V = B_{\leq \gamma}(e)$, so again, V is closed.

We now show V is C-definable. If $a \in V'$ for some maximal open sub-ball V' of V, with V' defined over $\mathcal{U}$ then by definition of V, V' cannot be defined over M; so V' has an $\operatorname{Aut}(\mathcal{U}/M)$-conjugate $V'' \neq V'$; but then $V' \cap V'' = \emptyset$, a contradiction as $a \in V''$ by invariance. Thus a is generic in the closed ball $V = B_{\leq \gamma}(e)$ over $\mathcal{U}$. Finally, V is unique with this property, so V is $\operatorname{Aut}(\mathcal{U}/C)$-invariant and hence C-definable. $\qquad\square$

In the next theorem, the base change results for stable domination in Chapter 4 are crucial.

THEOREM 10.7. *Let p be an $\operatorname{Aut}(\mathcal{U}/C)$-invariant $*$-type (so possibly in infinitely many variables). Suppose that for any model $M \supseteq C$ and $a \models p|M$ we have $\Gamma(M) = \Gamma(Ma)$. Then $p|C$ is stably dominated.*

PROOF. We shall show that if $C \subseteq M$ and $a \models p|M$ then $\operatorname{tp}(a/M)$ is stably dominated. For then by Theorem 4.9, $\operatorname{tp}(a/C)$ is stably dominated.

We first show how to replace a by a tuple of field elements. Choose a small model M with $C \subseteq M$. Let $a' := aa_1$ be a sequence such that $\operatorname{acl}(Ma')$ is a generic closed resolution of a over M. We choose a_1 to contain, for each lattice with code s in the sequence a, a basis of $\Lambda(s)$ which is generic over M. Note that $\operatorname{tp}(a'/M)$ has a canonical $\operatorname{Aut}(\mathcal{U}/M)$-invariant extension p' over $\mathcal{U}$ whose restriction to the a-variables is p. Put $A := \operatorname{acl}(Ma')$. Now, as in the paragraph before Lemma 10.5, let $k_{\operatorname{rep}}(A)$ be the subfield of

$k(A)$ consisting of res(0) and elements res(x) where $x \in A$ and $|x| = 1$. Let $\mathcal{B}, J$ be as in the definition of canonical open resolution of A. Let a_2 list J, and put $a^* := a'a_2$. Then acl(Ma^*) is a canonical open resolution of A. All elements in S in acl(Ma') are resolved, and $\Gamma(M) = \Gamma(Ma')$. Also, acl(Ma^*) is resolved. Let p^* be any Aut($\mathcal{U}/M$)-invariant extension over $\mathcal{U}$ of tp(a^*/M) whose restriction to the aa'-variables is p'. By omitting some elements of a^* if necessary, we may suppose that a^* is a sequence of field elements (with $a \in$ dcl(Ma^*)). Now suppose $a^* \models p^*|M'$ for some model $M' \supseteq M$. Then $k(Ma')$ and $k(M')$ are independent in VS$_{k,M}$. It follows that acl($M'a^*$) is a canonical open resolution of acl($M'a'$), and that $\Gamma(M'a^*) = \Gamma(M'a') = \Gamma(M')$.

We shall show that tp(a^*/M) is stably dominated. For then tp(a/M) is stably dominated by Proposition 3.32 (ii), and this suffices as described above. Put $A^* := $ acl(Ma^*) $\cap K$.

To show that tp(a^*/M) is stably dominated, we must show that if N is a model of finite transcendence degree over M, and $k(A^*)$ and $k(N)$ are independent over $k(M)$, then $a^* \underset{M}{\overset{g}{\smile}} N$; for this ensures, by Theorem 8.12, that tp($N/Mk(A^*)$) $\vdash$ tp(N/A^*). We prove this by induction on the transcendence degree n of N over M.

Suppose first that $n = 1$, that is, $N = $ acl(Mb) where $b \in K$. Then by Proposition 8.21, we have $A^* \underset{M}{\overset{g}{\smile}} N \leftrightarrow b \underset{M}{\overset{g}{\smile}} MA^*$, and there is no dependence on the choice of generating sequence. By Proposition 8.22, we may suppose b embeds into A^* over M. By Example 10.6, b is generic in a closed ball over M. This closed ball is in M-definable bijection with R, so we may suppose that $b \in R$. Since $k(Mb)$ and $k(A^*)$ are independent over $k(M)$, we have $b \underset{M}{\overset{g}{\smile}} A^*$, as required.

For the inductive step, suppose that $M \subset M' \subset N$ with trdeg(N/M') $= 1$. By induction, $a^* \underset{M}{\overset{g}{\smile}} M'$, and by Proposition 8.19, $k(A^*)$ and $k(M')$ are independent (in St$_M$ over $k(M)$). Hence,

$$\mathrm{trdeg}(k(M'a^*)/k(M')) = \mathrm{trdeg}(k(Ma^*)/k(M)).$$

The latter equals trdeg($k(Na^*)/k(N)$), by the assumption that $k(A^*)$ and $k(N)$ are independent over $k(M)$. Thus, $k(M'a^*)$ and $k(N)$ are independent over $k(M')$. It follows that, by the $n = 1$ case, $a^* \underset{M'}{\overset{g}{\smile}} N$. Hence, by transitivity of $\overset{g}{\smile}$, we have $a^* \underset{M}{\overset{g}{\smile}} N$ as required. $\qquad\square$

COROLLARY 10.8. *Let a be a unary sequence over C (possibly transfinite). Then $p := $ tp(a/C) is stably dominated if and only if it is orthogonal to Γ.*

PROOF. We may suppose $C = $ acl(C), as both conditions are unaffected by this. Suppose first that p is stably dominated. Then by Proposition 3.13, p has a unique Aut($\mathcal{U}/C$)-invariant extension $p|\mathcal{U}$. Let $M \supseteq C$ be a model. If $a \models p|M$, then $a \underset{C}{\overset{g}{\smile}} M$ (see Proposition 8.17). Thus, by uniqueness of sequential extensions, if $a \underset{C}{\overset{g}{\displaystyle\downarrow}} M$, then $a \models p|M$, a stably dominated type by Proposition 4.1. Hence, by Lemma 8.20, $\mathrm{tp}(a/M) \vdash \mathrm{tp}(a/M\Gamma(\mathcal{U}))$. It follows that $\Gamma(M) = \Gamma(Ma)$. Thus $\mathrm{tp}(a/C) \perp \Gamma$.

Conversely, suppose $p \perp \Gamma$. Let $p|\mathcal{U}$ be the $\mathrm{Aut}(\mathcal{U}/C)$-invariant extension given by sequential independence. Then for any model $M \supseteq C$, if $a \models p|M$ we have $a \underset{C}{\overset{g}{\displaystyle\downarrow}} M$, so $\Gamma(M) = \Gamma(Ma)$. It follows by Theorem 10.7 that p is stably dominated. $\hfill\square$

Observe that in the above setting, $\mathrm{tp}(a/C)$ is stably dominated if and only if $\mathrm{tp}(\mathrm{acl}(Ca)/C)$ is stably dominated (by Proposition 3.32 (iii) and Corollary 6.12). We use this freely.

LEMMA 10.9. *Let* $\mathrm{acl}(Ca) = \mathrm{acl}(Ca')$ *with* a, a' *both unary sequences over* C. *Suppose that* $\mathrm{tp}(a/C) \perp \Gamma$. *Then*

  (i) $a \underset{C}{\overset{g}{\displaystyle\downarrow}} B$ *if and only if* $a' \underset{C}{\overset{g}{\displaystyle\downarrow}} B$,

  (ii) $\mathrm{tp}(a'/C) \perp \Gamma$.

PROOF. We may suppose $C = \mathrm{acl}(C)$. Let $A := \mathrm{acl}(Ca)$. By Corollary 10.8, $\mathrm{tp}(A/C)$ is stably dominated. Hence, by Proposition 3.13, for any saturated model M there is a unique $\mathrm{Aut}(M/C)$-invariant extension of $\mathrm{tp}(A/C)$ to M.

(i) Suppose $a \underset{C}{\overset{g}{\displaystyle\downarrow}} B$. By Proposition 8.19, $A^{\mathrm{st}} \underset{C}{\displaystyle\downarrow} B^{\mathrm{st}}$ and hence by stable domination, the unique $\mathrm{Aut}(\mathcal{U}/C)$-invariant extension of $\mathrm{tp}(A/C)$ extends $\mathrm{tp}(A/B)$. There is $a'' \equiv_C a'$ with $a'' \underset{C}{\overset{g}{\displaystyle\downarrow}} B$ such that $\mathrm{tp}(\mathrm{acl}(Ca'')/B)$ extends to some $\mathrm{Aut}(\mathcal{U}/C)$-invariant extension of $\mathrm{tp}(A/C)$ (Corollary 8.15). Hence, by the above uniqueness, $a' \underset{C}{\overset{g}{\displaystyle\downarrow}} B$.

Now suppose $a' \underset{C}{\overset{g}{\displaystyle\downarrow}} B$. Since $A = \mathrm{acl}(Ca) = \mathrm{acl}(Ca')$, the same argument shows that $A \underset{C}{\overset{g}{\displaystyle\downarrow}} B$ via any generating sequence.

(ii) Suppose $a' \underset{C}{\overset{g}{\displaystyle\downarrow}} M$. Then by (i) $a \underset{C}{\overset{g}{\displaystyle\downarrow}} M$, so $\Gamma(Ma) = \Gamma(M)$. But $\Gamma(Ma) = \Gamma(Ma')$. Hence $\Gamma(Ma') = \Gamma(M)$. $\hfill\square$

DEFINITION 10.10. Let $A = \mathrm{acl}(Ca)$, with $\mathrm{tp}(a/C)$ unary. We write $\mathrm{tp}(A/C) \perp \Gamma$ if $\mathrm{tp}(a/C) \perp \Gamma$. By the last lemma, this definition is independent of the choice of acl-generating sequence a.

PROPOSITION 10.11. *Suppose* $\mathrm{tp}(A/C) \perp \Gamma$. *Then the following are equivalent.*

  (i) $\mathrm{St}_C(A) \underset{C}{\displaystyle\downarrow} \mathrm{St}_C(B)$.

  (ii) $A \underset{C}{\overset{d}{\displaystyle\downarrow}} B$.

  (iii) $A \underset{C}{\overset{g}{\displaystyle\downarrow}} B$ *via some generating sequence.*

  (iv) $A \underset{C}{\overset{g}{\displaystyle\downarrow}} B$.

  (v) $B \underset{C}{\overset{d}{\displaystyle\downarrow}} A$.

  (vi) $B \underset{C}{\overset{g}{\displaystyle\downarrow}} A$ *via some generating sequence.*

(vii) $B \underset{C}{\overset{g}{\perp}} A$.

PROOF. We may suppose $C = \mathrm{acl}(C)$. First, by Corollary 10.8, $\mathrm{tp}(A/C)$ is stably dominated.

The equivalence (i) $\Leftrightarrow$ (ii) is by the definition of stable domination, and (iii) $\Leftrightarrow$ (iv) comes from Lemma 10.9. We get (ii) $\Rightarrow$ (iii) from Proposition 8.17 (as there is a unary sequence a with $A = \mathrm{acl}(Ca)$); (iii) $\Rightarrow$ (i) is by Proposition 8.19.

The equivalence (ii) $\Leftrightarrow$ (v) is Proposition 3.8 (i). To prove (v) $\Rightarrow$ (vi), suppose $B \underset{C}{\overset{d}{\perp}} A$, and $B = \mathrm{acl}(Cb)$, b a unary sequence. There is an automorphism σ over C with $\sigma(b) \underset{C}{\overset{g}{\perp}} A$. By Lemma 8.19, we have

$$\mathrm{St}_C(\mathrm{acl}(C\sigma(b))) \underset{C}{\perp} \mathrm{St}_C(A),$$

which gives $\sigma(B) \underset{C}{\overset{d}{\perp}} A$. It follows that $B \equiv_A \sigma(B)$, so $b \underset{C}{\overset{g}{\perp}} A$, as required. Next, we show (vi) $\Rightarrow$ (i). For if $b \underset{C}{\overset{g}{\perp}} A$ then by Proposition 8.19 we have $\mathrm{St}_C(B) \underset{C}{\perp} \mathrm{St}_C(A)$, which gives (i) by symmetry of stable forking. Finally, (vii) $\Rightarrow$ (vi) trivially, and (vi) $\Rightarrow$ (vii) is immediate, for the first 6 conditions are equivalent, and in the implication (v) $\Rightarrow$ (vi), the choice of b was arbitrary. $\quad\square$

We now extend the resolution results further, in order to obtain 10.16 below.

LEMMA 10.12. *Suppose $A = \mathrm{acl}(A) \cap K$, and $\mathcal{B} \subset k$. Then $\mathrm{acl}(A \cup \mathcal{B}) \cap K = A$.*

PROOF. If $a \in \mathrm{acl}(A \cup \mathcal{B}) \cap K$, then by elimination of imaginaries in ACF, the set of conjugates of a over $A \cup \mathcal{B}$ is coded by some tuple a' from K. Now a' lies in some A-definable k-internal subset of K^m for some m, and it follows from Lemma 7.10 that $a' \in A$. $\quad\square$

LEMMA 10.13. *Suppose $C = \mathrm{acl}(C) \subseteq A = \mathrm{acl}(A)$, $\mathrm{tp}(A/C) \perp \Gamma$, and C is resolved, and that all elements of S_n in A are resolved and that A is finitely acl-generated over C. Let B be a canonical open resolution of A. Then $\mathrm{tp}(B/C) \perp \Gamma$.*

PROOF. In the notation before Lemma 10.5, let $\mathcal{B}$ be a transcendence basis of $k(A)$ over $k_{\mathrm{rep}}(A)$, and suppose $\mathcal{B} = \{\mathrm{res}(c) : c \in J\}$. Let $B := \mathrm{acl}(A \cup J)$. We may suppose that $\mathcal{B}$ is finite. There is a finite sequence a_1 from A_K with $A_K = \mathrm{acl}(Ca_1) \cap K$. Let a_2 be an enumeration of $\mathcal{B}$. Then as in Lemma 10.5, $A = \mathrm{acl}(Ca_1a_2)$. Clearly a_1a_2 is unary. By Lemma 10.9, $\mathrm{tp}(a_1a_2/C) \perp \Gamma$. Let M be a model, and suppose $a_1a_2J \underset{C}{\overset{g}{\perp}} M$. It suffices to show $\Gamma(M) = \Gamma(MB)$. Now $a_1a_2 \underset{C}{\overset{g}{\perp}} M$, so $\Gamma(M) = \Gamma(MA)$ since $\mathrm{tp}(a_1a_2/C) \perp \Gamma$.

We claim that $\mathrm{acl}(BM)$ is a canonical open resolution of $\mathrm{acl}(AM)$; for then, by Lemma 10.5, $\Gamma(MB) = \Gamma(MA)$, so $\Gamma(MB) = \Gamma(M)$. We require that $\mathcal{B}$ is a transcendence basis of $k(AM)$ over $k_{\mathrm{rep}}(AM)$; that is, $\mathcal{B}$ is algebraically independent over $\{\mathrm{res}(x) : x \in K \cap \mathrm{acl}(AM)\}$. By Lemma 10.12, $\mathrm{acl}(AM \cap K) =$

$\operatorname{acl}(A_K \cup M_K \cup B) \cap K = \operatorname{acl}(A_K M_K) \cap K$. Also, $a_2 \underset{A_K}{\overset{g}{\smile}} A_K M_K$, and it follows that the elements of a_2 are algebraically independent over $k_{\text{rep}}(AM)$. This yields the claim, and hence the lemma. $\qquad \square$

LEMMA 10.14. *Suppose $C = \operatorname{acl}(C)$ is resolved, and suppose $\operatorname{tp}(a/C) \perp \Gamma$. Then there is a resolved $\mathcal{L}_G$-structure B containing Ca, with $\operatorname{tp}(B/C) \perp \Gamma$.*

PROOF. Choose A to be a generic closed resolution of Ca, and B to be a canonical open resolution of A, and apply Remark 10.4 (iii) and Lemma 10.13. $\qquad \square$

PROPOSITION 10.15. *Let C be an $\mathcal{L}_G$-structure. Then*
  (i) *there is a resolved $\mathcal{L}_G$-structure $C' \supseteq C$ with $\Gamma(C) = \Gamma(C')$,*
  (ii) *there is a resolved $\mathcal{L}_G$-structure $C'' \supseteq C$ with $k(\operatorname{acl}(C)) = k(C'')$.*

PROOF. We may suppose that $C = \operatorname{acl}(C)$.
  (i) Apply Remark 10.4 (ii) and Lemma 10.5.
  (ii) It suffices to show that we may resolve any set coded by an element of $S_n \cup T_n$ without increasing $k(C)$.

Let s be a code for a lattice of S_n. Then $\operatorname{red}(s)$ is an n-dimensional vector space over k. We show that we can add elements to C to get a basis of $\operatorname{red}(s)$ to C without increasing $k(C)$. Let $U_1, \ldots, U_r$ be a maximal k-linearly independent subset of $\operatorname{red}(s)$ with codes in C, and suppose that $r < n$. Choose U generic in $\operatorname{red}(s)$ over C. We may suppose that $k(C) \neq k(C^{\ulcorner}U^{\urcorner})$, since otherwise we may add $^{\ulcorner}U^{\urcorner}$ to C and so get a larger k-linearly independent set in $\operatorname{red}(s)$. Thus there is a C-definable partial function f with $f(^{\ulcorner}U^{\urcorner}) \in k \setminus k(C)$. Let X equal

$$\{U' \in \operatorname{red}(s) \colon {}^{\ulcorner}U'^{\urcorner} \in \operatorname{dom}(f) \text{ and}$$

$$U_1, \ldots, U_r, U' \text{ are linearly independent over } k\}.$$

Then $\{f(^{\ulcorner}U'^{\urcorner}) \colon U' \in X\}$ is an infinite definable subset of k, so as k is strongly minimal, there is $U' \in X$ with $f(^{\ulcorner}U'^{\urcorner}) \in k(C)$. Thus, we may add $^{\ulcorner}U'^{\urcorner}$ to C without increasing $k(C)$.

It remains to show that if $U \in \operatorname{red}(s)$ with $^{\ulcorner}U^{\urcorner} \in \operatorname{dcl}(C)$, then there is $a \in U \cap K^n$ with $k(C) = k(Ca)$. We prove, by induction on n, that this holds for any coset in K^n of an element with code in T_n (that is, any $U + x$ where $^{\ulcorner}U^{\urcorner} \in T_n$ and $x \in K^n$). So let U be such a coset, and let U_1 be its projection to the first coordinate. Then U_1 is an open ball, so if a_1 is chosen generically in U_1 over C, then $k(C) = k(Ca_1)$ by Lemma 7.25. Now $U^1 := \{x \in K^{n-1} \colon (a_1, x) \in U\}$ is a coset of an element with code in T_{n-1}. Hence, by induction, there is $b \in U^1$ with $k(Ca_1) = k(Ca_1 b)$. Now $(a_1, b) \in U$ and $k(C) = k(Ca_1 b)$, as required. $\qquad \square$

These results on resolutions will be extended further in Chapter 11, where prime resolutions are considered.

The next proposition indicates another sense in which all instability arises from interaction with the value group.

PROPOSITION 10.16. *Let $(a_i : i < \omega)$ be an indiscernible sequence over $C \cup \Gamma(\mathcal{U})$. Then $\{a_i : i < \omega\}$ is an indiscernible set over $C \cup \Gamma(\mathcal{U})$.*

PROOF. We may assume $C = \operatorname{acl}(C)$. In the proof below, we often just write Γ for $\Gamma(\mathcal{U})$.

Since $(a_i : i < \omega)$ is indiscernible over $C \cup \Gamma$, we have

$$\Gamma(Ca_{i_1} \ldots a_{i_n}) = \Gamma(Ca_{j_1} \ldots a_{j_n})$$

for any n and any $i_1 < \cdots < i_n, j_1 < \cdots < j_n$. By adding all such sets to C, we may arrange that for all such n and $i_1 < \cdots < i_n$ we have

$$\Gamma(Ca_{i_1} \ldots a_{i_n}) = \Gamma(C). \tag{1}$$

Also, for each n there is a fixed $C_n \supset C$ such that for all $i_1 < \cdots < i_n < j_1 < \cdots < j_n$ (chosen in ω) we have $\operatorname{acl}(Ca_{i_1} \ldots a_{i_n}) \cap \operatorname{acl}(Ca_{j_1} \ldots a_{j_n}) = C_n$. By expanding C we may suppose that $C_n = C$ for each $n \in \omega$; that is, for all $n < \omega$ and $i_1 < \cdots < i_n < j_1 < \cdots < j_n$,

$$\operatorname{acl}(Ca_{i_1} \ldots a_{i_n}) \cap \operatorname{acl}(Ca_{j_1} \ldots a_{j_n}) = C. \tag{2}$$

We preserve also

$$(a_i : i < \omega) \text{ is an indiscernible sequence over } C \cup \Gamma. \tag{3}$$

We will show that $\operatorname{tp}(a/C) \perp \Gamma$ for each finite subsequence a of $(a_i : i < \omega)$. We first argue that this suffices, so suppose it holds. Let $a_i' := \operatorname{VS}_{k,C}(a_i)$ for each $i < \omega$. Then $(a_i' : i < \omega)$ is an indiscernible sequence in the stable structure $\operatorname{VS}_{k,C}$, so is an indiscernible set. It follows from this and (2) that, for any n and $i_1 < \cdots < i_n$ and $j_1 < \cdots < j_n$ with $\{i_1, \ldots i_n\} \cap \{j_1, \ldots, j_n\} = \emptyset$, we have $\operatorname{acl}(Ca_{i_1}' \ldots a_{i_n}') \cap \operatorname{acl}(Ca_{j_1}' \ldots a_{j_n}') = C$. It follows (as in Proposition 3.16) that $\{a_i' : i \in \omega\}$ is an independent set in $\operatorname{VS}_{k,C}$. Using orthogonality to Γ and Proposition 6.10 (ii), we obtain

$$\operatorname{dcl}(a_{i_1}', \ldots a_{i_n}') = \operatorname{VS}_{k,C}(a_{i_1} \ldots a_{i_n}) = \operatorname{VS}_{k,C}(a_{i_1} \ldots a_{i_n} \Gamma),$$

(the second equality holds as any definable function from Γ to the stable structure $\operatorname{VS}_{k,C}$ has finite range). Hence, by indiscernibility, for any $i, j \in \omega \setminus \{i_1, \ldots, i_n\}$,

$$\operatorname{tp}(a_i' / \operatorname{VS}_{k,C}(a_{i_1} \ldots a_{i_n} \Gamma)) = \operatorname{tp}(a_j' / \operatorname{VS}_{k,C}(a_{i_1} \ldots a_{i_n} \Gamma)).$$

It follows as in Remark 3.7 that

$$\operatorname{tp}(a_i / \operatorname{VS}_{k,C}(a_{i_1} \ldots a_{i_n} \Gamma)) = \operatorname{tp}(a_j / \operatorname{VS}_{k,C}(a_{i_1} \ldots a_{i_n} \Gamma)),$$

and hence by Proposition 10.11 that

$$\operatorname{tp}(a_i / Ca_{i_1} \ldots a_{i_n} \Gamma) = \operatorname{tp}(a_j / Ca_{i_1} \ldots a_{i_n} \Gamma).$$

This yields the indiscernibility claimed in the theorem.

It remains to prove that $\mathrm{tp}(a/C) \perp \Gamma$ for each such a. For this, it suffices to show

CLAIM. Suppose that $C' = \mathrm{acl}(C') \supseteq C$, and (1'), (2') and (3') hold, where these are (1), (2), and (3) with the base C replaced by C'. Suppose also b is chosen in K so that for each n and $i_1 < \cdots < i_n$ we have $a_{i_1} \ldots a_{i_n} \underset{C'}{\overset{g}{\downarrow}} b$. Then (1''), (2''), and (3'') hold, where these are (1), (2), and (3) respectively, with the base C replaced by $C'' = \mathrm{acl}(C'b)$.

Given the claim, suppose M is a model with $a \underset{C}{\overset{g}{\downarrow}} M$ for each finite subsequence a of $(a_i : i \in \omega)$. We may suppose $M = \mathrm{acl}(Cb)$ for some finite sequence b of field elements. By applying the claim repeatedly, we find $\Gamma(M) = \Gamma(Ma)$ for each finite subsequence a of $(a_i : i < \omega)$. This proves $\mathrm{tp}(a/C) \perp \Gamma$.

PROOF OF CLAIM. We first prove (1''). This certainly holds if b is a generic element of a closed 1-torsor over C'. For then $\mathrm{tp}(b/C') \perp \Gamma$, so $b \underset{C'}{\overset{g}{\downarrow}} \{a_i : i < \omega\}$, by Proposition 10.11. Hence,

$$\Gamma(C') = \Gamma(C'a_i : i < \omega) = \Gamma(C'ba_i : i < \omega) = \Gamma(C'b). \qquad (4)$$

It remains to consider the case when b is a generic element of an open 1-torsor U over C' (or of an ∞-definable 1-torsor — this case is similar). First, if not all elements of U have the same type over C', then, as any C'-definable subset of U is a finite union of C'-definable Swiss cheeses, some proper subtorsor of U is algebraic over C'. Hence $\Gamma(C'b) \neq \Gamma(C')$ by Lemma 7.27, so $\mathrm{rk}_{\mathbb{Q}}(\Gamma(C'b)) = \mathrm{rk}_{\mathbb{Q}}(\Gamma(C')) + 1$ by Lemma 7.11. In this case (1'') follows from (1') by rank considerations (Lemma 7.11). Thus, we assume all elements of U have the same type over C'. We may suppose they do not all have the same type over $C' \cup \{a_i : i < \omega\}$, since otherwise (4) is again valid. Hence, there are $i_1, \ldots, i_n$ and an $a_{i_1} \ldots a_{i_n}$-definable proper subtorsor $U(a_{i_1}, \ldots, a_{i_n})$ of U which intersects the $\mathrm{tp}(b/C')$ non-trivially. By (1'), the radius of $U(a_{i_1}, \ldots, a_{i_n})$ is in C', and so also is the radius of $U(a_{i_1'}, \ldots, a_{i_n'})$ for any other $a_{i_1'}, \ldots, a_{i_n'}$ from the sequence. So we have a uniform family of definable subtorsors of an open 1-torsor, all of the same radius. The distance between two such torsors is a C'-definable function of $2n$ variables, hence by (1') has constant value, say γ, and $\gamma \in C'$. But then all the torsors $U(a_{i_1'}, \ldots, a_{i_n'})$, and in particular, $U(a_{i_1}, \ldots, a_{i_n})$ are contained in a C'-definable closed 1-torsor properly contained in U. This contradicts the assumption that all elements of U have the same 1-type over C'.

To see (2''), suppose $e \in \mathrm{acl}(C'ba_{i_1} \ldots a_{i_n}) \cap \mathrm{acl}(C'ba_{j_1} \ldots a_{j_n})$, where $i_1 < \cdots < i_n < j_1 < \cdots < j_n$. By increasing n if necessary we may suppose the number m of conjugates of e over $C'ba_{i_1} \ldots a_{i_n}$ is as small as possible. Then there is a definable function f on $\mathrm{tp}(b/C')$ taking b to a code for the m-set X consisting of realisations of $\mathrm{tp}(e/C'ba_{i_1} \ldots a_{i_n})$. By minimality of m, X

is also a complete type over $C'ba_{j_1} \ldots a_{j_n}$; otherwise e has fewer than m conjugates over $C'ba_{i_1} \ldots a_{i_n} a_{j_1} \ldots a_{j_n}$. Hence by indiscernibility (over $C'b$) X is a complete type over any such $C'ba_{\ell_1} \ldots a_{\ell_n}$. Thus, $\ulcorner X \urcorner \in \mathrm{dcl}(C'ba_{\ell_1} \ldots a_{\ell_n})$ for any $\ell_1 < \cdots < \ell_n$. Thus, there is a definable function f on $\mathrm{tp}(b/C')$ with $f(b) = \ulcorner X \urcorner$, and $\ulcorner f \urcorner \in \mathrm{acl}(C'a_{i_1} \ldots a_{i_n}) \cap \mathrm{acl}(C'a_{j_1} \ldots a_{j_n}) = \mathrm{acl}(C')$, so $e \in \mathrm{acl}(C'b)$, as required.

Finally, for $(3'')$, first observe that by uniqueness of sequentially independent extensions (Corollary 8.13), $(a_i : i < \omega)$ is an indiscernible sequence over C''. Condition $(3'')$ now follows easily from $(1'')$. $\qquad \square$

CHAPTER 11

OPACITY AND PRIME RESOLUTIONS

In this chapter we extend the results on resolutions from Chapter 10. In particular, we show (Theorem 11.14) that over an algebraically closed base in the field sorts, any finite tuple has a unique minimal atomic 'prime' resolution, which does not extend either the value group or the residue field. We talk of 'prime resolutions' rather than 'prime models' since the resolution may not be a model of ACVF, as the valuation may be trivial on it.

The first few lemmas (11.3–11.12) are proved in complete generality. We suppose M is a sufficiently saturated structure, and emphasise that M consists of elements of the home sort, that is, we are not identifying M with M^{eq}. Unless otherwise specified, parameter sets and tuples below live in M^{eq}. In the application to ACVF, M will be the sort K of field elements. We work in the model M, over a parameter base C.

DEFINITION 11.1. Let A be a substructure of M^{eq}. A *pre-resolution* of A is a substructure D of M^{eq} with $A \subseteq \mathrm{acl}(D \cap M)$.

A *resolution* of A is an algebraically closed structure $D \subseteq M$ such that $A \subseteq \mathrm{dcl}(D)$. A resolution D is *prime* over A if D embeds over A into any other resolution.

Usually, A will have the form Ca where $C \subseteq M$, and a is a finite sequence of imaginaries.

In the intended application to ACVF, the sequence a may be taken from $S \cup T$. We shall exploit the fact that S_n and T_n can be viewed a coset spaces of groups of upper triangular matrices. Such groups have sequences of normal subgroups whose quotients are strongly minimal (the additive or multiplicative group of the field, in the case of S_n). This leads us to the following sequence of definitions and lemmas.

DEFINITION 11.2. Let E be a definable equivalence relation on a definable set D (in M). We say E is *opaque* if, for any definable $Z \subset D$, there are Z' (a union of E-classes) and Z'' (contained in the union of finitely many E-classes) such that $Z = Z' \cup Z''$.

For example, in ACVF, the partition of R into cosets of $\mathcal{M}$ is an opaque equivalence relation; the same holds for the partition of any closed ball into open sub-balls of the same radius.

LEMMA 11.3. *Let E be a C-definable opaque equivalence relation on D, and let $a = b/E$ be a class of E. Then either $a \in \mathrm{acl}(C)$, or $\mathrm{tp}(b/Ca)$ is isolated, where the isolating formula $\varphi(y)$ states that $y/E = a$.*

The terms 'isolated type' and likewise, below, 'atomic' are interpreted with respect to $\mathrm{Th}(M)$. Thus, $\mathrm{tp}(b/Ca)$ might be isolated but not realised in some resolution of Ca (e.g., if the latter is trivially valued, in the ACVF context).

PROOF. If $\varphi(y)$ does not isolate a complete type over Ca, there is a formula $\psi(y, x)$ over C such that $\psi(y, a)$ defines a proper non-empty subset of a. Let Z be defined by $\psi(y, y/E)$. Then Z meets the class a in a proper non-empty subset. Since E is opaque, there are just finitely many such classes. Hence $a \in \mathrm{acl}(C)$.   □

LEMMA 11.4. *Let $a_0, \ldots, a_{N-1}$ be a sequence of (imaginary) elements. Assume that $a_n = b_n/E_n$ for each $n \in \{0, \ldots, N-1\}$, where E_n is an opaque equivalence relation defined over $A_n := C \cup \{a_j : j < n\}$. Define B_n by downward recursion: $B_N = \emptyset$ and $B_n := B_{n+1} \cup \{b_n\}$ if $a_n \notin \mathrm{acl}(B_{n+1} \cup A_n)$, and $B_n := B_{n+1}$ otherwise. Let $I := \{n : b_n \in B_0\}$ (which equals $\{n : a_n \notin \mathrm{acl}(B_{n+1} \cup A_n)\}$). Then*

(i) $A_N \subset \mathrm{acl}(B_0 \cup C)$,
(ii) *if $n \in I$ then $\mathrm{tp}(b_n/A_N \cup B_{n+1})$ is isolated.*
(iii) B_0 *is atomic over A_N.*

PROOF. (i) We have for each n

$$a_n \in \mathrm{acl}(B_n \cup A_n). \tag{1}$$

If $n \notin I$, this follows by definition as $a_n \in \mathrm{acl}(B_{n+1} \cup A_n) \subset \mathrm{acl}(B_n \cup A_n)$. If $n \in I$, then $a_n = b_n/E_n \in \mathrm{acl}(B_n)$.

Also, for each k, we have

$$\text{if } n \geq k \text{ then } a_n \in \mathrm{acl}(B_k \cup A_k). \tag{2}$$

This is proved by induction on $n \geq k$. If $n \geq k$ then by (1) (as $B_k \supseteq B_n$) we have $a_n \in \mathrm{acl}(B_k \cup A_n)$. By induction, $A_n \subset \mathrm{acl}(B_k \cup A_k)$, so $a_n \in \mathrm{acl}(B_k \cup A_k)$.

Applying (2) with $k = 0$ we have (i).

(ii) Let $n \in I$. We apply Lemma 11.3 with C replaced by $A_n \cup B_{n+1}$ to obtain that $\mathrm{tp}(b_n/B_{n+1} \cup A_n \cup \{a_n\})$ is isolated. By (2), $A_N \subset \mathrm{acl}(B_{n+1} \cup A_n \cup \{a_n\})$. Hence (ii) holds.

(iii) This follows immediately from (ii).   □

DEFINITION 11.5. (i) We say that $\mathrm{tp}(a/C)$ is *opaquely layered* (or a *is opaquely layered over C*) if there exist $a_0, \ldots, a_{N-1}, E_0, \ldots, E_{N-1}$ satisfying the hypotheses of Lemma 11.4 such that $\mathrm{dcl}(C, a) = \mathrm{dcl}(C, a_0, \ldots, a_{N-1})$.

(ii) Let E be a definable equivalence relation on a definable set D. Then (D, E) is *opaquely layered over* C if D, E are defined over C and for each $a \in D$, the imaginary a/E is opaquely layered over C. We say (D, E) is *opaquely layered everywhere* if for any C' and any pair (D', E') definable over C' and definably isomorphic (over some C'') to (D, E), the pair (D', E') is opaquely layered over C'.

(iii) Let G be a definable group, and F a definable subgroup. Then G/F is *opaquely layered over* C (respectively, *opaquely layered everywhere*, respectively, *opaque*) if the pair (G, E) is, where Exy is the equivalence relation $xF = yF$.

Note that opacity is preserved under definable bijections, so 'opaque' implies 'opaquely layered everywhere'. Also, if $\mathrm{tp}(a/C)$ is opaquely layered and $C \subset B$, then $\mathrm{tp}(a/B)$ is opaquely layered.

LEMMA 11.6. *Suppose* $\mathrm{tp}(a/C)$ *and* $\mathrm{tp}(b/Ca)$ *are opaquely layered. Then* $\mathrm{tp}(ab/C)$ *is opaquely layered.*

PROOF. Opaque layering of $\mathrm{tp}(ab/C)$ is witnessed by concatenating a sequence for $\mathrm{tp}(a/C)$ (as in Lemma 11.4) and a sequence for $\mathrm{tp}(b/Ca)$. ☐

PROPOSITION 11.7. *Suppose that* $C \subset M$ *and* a *is a finite sequence of imaginaries, and suppose that* $\mathrm{tp}(a/C)$ *is opaquely layered. Let* $a_0, E_0, \ldots, a_{N-1}, E_{N-1}$ *witness the opaque layering. Then* Ca *has a pre-resolution* D *which is atomic over* Ca *and embeds into* D' *for any* D' *which contains* C *and an element of each class* E_i-*class* a_i. *The embedding can be taken to be an elementary map.*

PROOF. Let I and b_i $(i \in I)$ and B_n $(n < N)$ be as in Lemma 11.4. Put $D := B_0 \cup C$. By Lemma 11.4, D is atomic over Ca, and $Ca \subseteq \mathrm{acl}(D)$.

For the second part, let D' be a resolution of a over C which contains an element b'_i of each E_i-class a_i. By part (ii) of Lemma 11.4, arguing inductively, the embedding property holds. More precisely, adopt the notation A_n, B_n of 11.4, and suppose an embedding $f_{n+1} : B_{n+1} \to D'$ has been defined over Ca. If $a_n \in \mathrm{acl}(B_{n+1} \cup A_n)$, then $B_n = B_{n+1}$ and we may put $f_n := f_{n+1}$. Otherwise, all elements of the equivalence class a_n have the same type over CaB_{n+1}, so we may extend f_{n+1} to f_n by putting $f_n(b_n) = b'_n$. Finally, f_0 is elementary, so extends to an embedding $D \to D'$. ☐

The following corollary applies for instance to $\mathrm{Th}(\mathbb{Q}_p)$ (taken in the $\mathcal{G}$-sorts.)

COROLLARY 11.8. *Assume any definably closed subset of* M *is an elementary submodel of* M. *Let* $C \subset M$ *and* a *a finite sequence of imaginaries, and suppose that* $\mathrm{tp}(a/C)$ *is opaquely layered. Then* Ca *has an atomic pre-resolution that embeds into any pre-resolution; the embedding can be taken to be an elementary map.* ☐

Define a dcl-*resolution* of Ca to be a subset D of M with $Ca \subseteq \text{dcl}(D)$. Note that (ii) below is true for any field; (i) holds for $\text{Th}(\mathbb{Q}_p)$ for instance, as well as for ACVF if one works over a nontrivially valued field.

COROLLARY 11.9. *Assume:*

(i) *any algebraically closed subset of M is an elementary submodel of M.*

(ii) *M admits elimination of finite imaginaries; i.e., any finite set of tuples of M is coded by elements of M.*

Let $C \subset M$ and let a be a finite sequence of imaginaries, and suppose that $\text{tp}(a/C)$ is opaquely layered. Let $a_0, E_0, \ldots, a_{N-1}, E_{N-1}$ witness the opaque layering. Then Ca has an atomic dcl-resolution B. Furthermore, B embeds into any $B' = \text{dcl}(B')$ which contains C and an element of each class E_i-class a_i.

Assume in addition that whenever an E_i-class Q is defined over a set C' and isolates a type q over C', this type is stationary. Then $\text{tp}(B/C) \vdash \text{tp}(B/\text{acl}(C))$.

PROOF. Let D be as in Proposition 11.7; so $Ca \subseteq \text{acl}(D \cap M)$, and D/Ca is atomic. By construction D is obtained from C by adding finitely many elements of M; at each stage one adds an element of an E_i-class which isolates a type over Ca and the preceding elements; by assumption, this type is stationary. It follows that $\text{tp}(D/Ca)$ is stationary.

Now $\text{acl}(D) \cap M$ is an elementary submodel of M; so $a \in \text{dcl}(e)$ for some $e \in \text{acl}(D) \cap M$. The orbit of e over Da is finite, hence coded by some tuple e' from M. We have $e' \in \text{dcl}(Da)$, and $a \in \text{dcl}(De')$. Let $B = De'$. Then B is a dcl-resolution of Ca, and all the conditions hold: $\text{tp}(e'/Da)$ is isolated and stationary, since $e' \in \text{dcl}(Da)$, so $\text{tp}(B/Ca)$ is isolated and stationary; and if $B' = \text{dcl}(B')$ contains C and an element of each class E_i-class a_i, then $D \subseteq B'$ and $a \in E'$, so $e' \in \text{dcl}(Da) \subseteq B'$. $\square$

LEMMA 11.10. *Let G be a C-definable group, and N, H, F be C-definable subgroups. Assume that N is normal, $N \cap H = \{1\}$, $NH = G$. Also suppose $F = N_F H_F$, where $K_F := K \cap F$ for any $K \le G$. Suppose that the coset space H/H_F is opaquely layered over C, and for each $h \in H$, $N/(N \cap F^h)$ is opaquely layered over $C \cup \{\ulcorner hH_F \urcorner\}$. Then G/F is opaquely layered over C.*

PROOF. Let $g \in G$. Then there are unique $n \in N$ and $h \in H$ with $g = nh$. By Lemma 11.6 and our assumptions, it suffices to show that

$$\text{dcl}(C, \ulcorner nhF \urcorner) = \text{dcl}(C, \ulcorner hH_F \urcorner, \ulcorner n(N \cap F^h) \urcorner),$$

where $F^h = hFh^{-1}$.

For the containment $\supseteq$, suppose that $nhF = n'h'F$, with $n, n' \in N$ and $h, h' \in H$. We must show $hH_F = h'H_F$ and $n(N \cap F^h)$ Then, first, $h' = hf''$ for some $f'' \in H_F$: indeed, $n'h' = nhf$ for some $f \in F$, and $f = n''f''$ where $n'' \in N_F$ and $f'' \in H_F$, hence $n'h' = nhn''f'' = nhn''h^{-1}hf''$, so $Nh' = Nhf''$. As $h', hf'' \in H$ and reduction modulo N is injective on H, we have $h' = hf''$. Thus, $hH_F = h'H_F$. Hence $\ulcorner hH_F \urcorner \in \text{dcl}(C, \ulcorner nhF \urcorner)$. Also,

$nhF = n'h'F = n'hf''F = n'hF$, so $nF^h = n'F^h$, and hence $n'(N \cap F^h) = n(N \cap F^h)$. As $h' = hf''$, $F^h = F^{h'}$. So $\ulcorner n(N \cap F^h)\urcorner \in \mathrm{dcl}(C, \ulcorner nhF\urcorner)$.

For the containment $\subseteq$, suppose that $hH_F = h'H_F$ and $n(N \cap F^h) = n'(N \cap F^{h'})$, where $n, n' \in N$ and $h, h' \in H$. The first equality yields that $hF = h'F$ so $n'hF = n'h'F$ and also $Fh^{-1} = Fh'^{-1}$; so $F^h = F^{h'}$, and $n(N \cap F^h) = n'(N \cap F^h)$ (by the second inequality). Hence, $n^{-1}n' \in F^h$, so $n'hF = nhF$, as required. $\square$

COROLLARY 11.11. *Let* $\{1\} = G_0 \leq G_1 \leq \cdots \leq G_N = G$ *and* $H_i \leq G_{i+1}$ *(for* $i = 0, \ldots N - 1$*) be* $\emptyset$*-definable groups, with* G_i *normal in* G, $G_i \cap H_i = \{1\}$ *and* $G_i H_i = G_{i+1}$ *for each* i. *Let* F *be a definable subgroup of* G *such that for each* i, $G_{i+1} \cap F = (G_i \cap F)(H_i \cap F)$ *and* $H_i/(H_i \cap F)$ *is opaquely layered everywhere. Then* G/F *is opaquely layered everywhere.*

PROOF. This follows from Lemma 11.10 and induction on i. Observe that if $G_i/(G_i \cap F)$ is opaquely layered everywhere, then by normality of G_i, and since opaque layeredness everywhere is preserved by definable bijections, $G_i/(G_i \cap F^h)$ is opaquely layered everywhere for each $h \in G$. $\square$

In the next lemma, by 'group scheme' we mean a system of polynomial equations, viewed as a formula, which for any integral domain A defines a subgroup of $B_n(A)$. Here, B_n is the group scheme such that $B_n(A)$ is the group of upper triangular matrices over A which are invertible in $\mathrm{GL}_n(A)$. As usual, K denotes a large algebraically closed field. We denote by $(e_1, \ldots, e_n)$ the standard basis for K^n.

LEMMA 11.12. *Fix* n, *and let* $N = \binom{n+1}{2}$.

(i) *The group scheme* B_n *has, for each* $i = 0, \ldots, N$, *group subschemes* G_i, H_i *over* $\mathbb{Z}$ *such that for any subring* A *of* K, *the following hold:* $G_{i+1}(A) = G_i(A)H_i(A)$, $G_i(A) \cap H_i(A) = \{1\}$, H_i *is isomorphic to* $(A, +)$ *or the group of multiplicative units of* A, $G_i(A)$ *is normal in* $B_n(A)$, $G_0 = \{1\}$, *and* $G_N = B_n$.

(ii) *For each* i, *if* $a \in G_i$, $b \in H_i$, *and* ab *fixes* e_n *(acting by left multiplication) then* a *and* b *both fix* e_n.

PROOF. The scheme B_n is defined so that $B_n(K)$ is the stabiliser of a maximal flag

$$\{0\} = V_0 \subset V_1 \subset \cdots \subset V_{n-1} \subset V_n = K^n;$$

here V_i is the space spanned by $\{e_1, \ldots, e_i\}$. For $0 \leq j \leq i \leq n$, let $G_{ij}(K)$ be the subgroup of $B_n(K)$ consisting of elements x such that x acts as the identity on V_{i-1}, and $x - I$ maps V_i into V_j. Then $G_{1,1} = B_n$, $G_{i,0} = G_{i+1,i+1}$, $G_{n,0} = \{1\}$, and

$$G_{n,0} < G_{n,1} < \cdots < G_{n,n} = G_{n-1,0} < \cdots < G_{i,0} < \cdots < G_{i,i}$$
$$= G_{i-1,0} < \cdots < G_{1,1} = B_n.$$

The G_{ij} are all normal in B_n since they are defined in terms of the flag. Also, it is easy to see that $G_{i,i} = G_{i,i-1} \trianglelefteq H_{i,i-1}$, where $H_{i,i-1}$ is the group of maps fixing each e_j ($j \neq i$) with eigenvector e_i (so $H_{i,i-1}(A)$ is isomorphic to $G_m(A)$, the group of units of A). Likewise, if $j < i$ then $G_{ij} = G_{i,j-1}H_{i,j-1}$, where $H_{i,j-1}(A)$ is isomorphic to $(A, +)$, and consists of elements fixing all e_k with $k \neq i$ and with $e_i \mapsto e_i + te_j$ ($t \in A$).

In terms of matrices, $G_{ii} = G_{i-1,0}$ and G_{ij} (for $j < i$) is the subset of $B_n(A)$ consisting of matrices which have top left $(i - 1) \times (i - 1)$-minor equal to I_{i-1}, and its ith column is the transpose of $(a_1, \ldots, a_n) \in A^n$ where $a_i = 1$ and $a_k = 0$ for $k > j$ with $k \neq i$. Also, $H_{i,i-1}$ is the group of invertible diagonal matrices in $B_n(A)$ with ones in all diagonal entries except the (i, i)-entry. For $j < i - 1$, H_{ij} has ones on the diagonal, any element of A in the (j, i)-position, and zeros elsewhere. It is easily verified that these are all groups.

(ii) If $i < n$ then all elements of H_{ij} fix e_n. Thus, we may suppose that $c = ab \in G_{nj}$ for some j. Now if c fixes e_n, then $c = 1$, so $a = b = 1$.   □

From now on, we revert to ACVF, so K denotes an algebraically closed valued field. Recall, the notation $B_{n,m}(R)$ from Chapter 7. In particular, $B_{n,n}(R)$ is the inverse image under the natural map $B_n(R) \to B_n(k)$ of the stabiliser in $B_n(k)$ of the standard basis vector e_n of k^n.

LEMMA 11.13. *Let K be an algebraically closed valued field.*
  (i) *The groups K/R and $K/\mathcal{M}$ (under addition) are opaque.*
  (ii) *If V is the group of units of R, then the multiplicative groups K^*/V and $K^*/(1 + \mathcal{M})$ are opaque.*
  (iii) *Let $G = B_n(K)$, $F = B_n(R)$ and $F' = B_{n,n}(R)$. Then G/F and G/F' are both opaquely layered everywhere.*

PROOF. (i) Every definable subset of K is a finite Boolean combination of balls. Thus, it suffices to verify that if U is a ball, then, for all but finitely many cosets $a + R$ of R, $a + R$ is either contained in U or disjoint from U (and likewise for $\mathcal{M}$). This is obvious.

(ii) This is similar to (i).

(iii) This follows by (i), (ii), Corollary 11.11 and Lemma 11.12. For the case of G/F, 11.12 provides a sequence $\{1\} = G_0 < \cdots < G_N = G$ as in 11.11, and the corresponding groups H_i. By the description above in terms of matrices of the H_{ij}, the groups $H_i/(H_i \cap F)$ and $H_i/(H_i \cap F')$ have the form described in (i) and (ii). Since 11.12 applies both to $G_i(K)$ and $G_i(R)$, the semidirect product decomposition required in 11.11 is clear for F.

We must also verify that for each i, $G_{i+1} \cap F' = (G_i \cap F')(H_i \cap F')$. Suppose $c \in G_{i+1} \cap F'$. Then $c = ab$ for some $a \in G_i \cap F$ and $b \in H_i \cap F$. Applying 11.12 for the field k, we see that this remains true in the reduction modulo $\mathcal{M}$. Hence the reduction of ab fixes the reduction of e_n. Thus the reductions of a and b both fix the reduction of e_n, by the last part of 11.12. It follows that $a \in G_i \cap F'$ and $b \in H_i \cap F'$, as required.   □

THEOREM 11.14. *Let C be a subfield of the algebraically closed valued field K, and let e be a finite set of imaginaries. Then Ce admits a resolution D which* *is minimal, prime and atomic over Ce. Up to isomorphism over Ce, D is the unique prime resolution of Ce. Also, $k(D) = k(\mathrm{acl}(Ce))$ and $\Gamma(D) = \Gamma(Ce)$.*

PROOF. We first prove existence of an atomic prime resolution. The easy case is when e has a trivially valued resolution B over C. In this case, by Lemma 7.5, the only B-definable lattice is R^n. Since R^n is resolved in C, we reduce to the case when $e = (e_1, \ldots, e_n)$ is a sequence of elements of k. Now e is opaquely layered over C via e_i, E_i, where E_i is just the equivalence relation $x + \mathcal{M} = y + \mathcal{M}$. Let D be the atomic resolution of Ce over C given by Proposition 11.7. Let B' be any resolution of e over C. To prove primality, we must show that B' satisfies the last condition in Proposition 11.7. The quantifier elimination of [12, Theorem 2.1.1(iii)] with sorts K, k, Γ yields the following, which is what is required: if $e_i \in k$ and $e_i \in \mathrm{acl}(B')$, then $\mathrm{acl}(B')$ contains a field element x with $\mathrm{res}(x) = e_i$.

Now suppose that every resolution of e over C is non-trivially valued. The tuple of imaginaries e has the same definable closure over C as some pair (a, b) where $a \in S_n$ and $b \in T_m$ for some n, m. We identify S_n with $B_n(K)/B_n(R)$. In Chapter 7 we identified T_m with $\bigcup_{i=1}^{\ell} B_m(K)/B_{m,\ell}(R)$. However, if $t \in B_m(K)/B_{m,\ell}(R)$ for $\ell < m$, then t may be identified with a pair (s', t') where $s' \in B_m(K)/B_m(R) = S_m$, and $t' \in B_\ell(K)/B_{\ell,\ell}(R)$; here, $t \in \mathrm{red}(s')$, and the coset t' consists of the $\ell \times \ell$ top left minors of the matrices in the coset t. Thus, after adjusting a and b if necessary, we may suppose that $a \in B_n(K)/B_n(R)$, and $b \in B_m(K)/B_{m,m}(R)$. By Lemma 11.13, $B_n(K)/B_n(R)$ and $B_m(K)/B_{m,m}(R)$ are opaquely layered, so $\mathrm{tp}(a/C)$ and $\mathrm{tp}(b/C)$ are opaquely layered (so $\mathrm{tp}(b/Ca)$ is). Hence, by Lemma 11.6, $\mathrm{tp}(ab/C)$ is opaquely layered.

By Proposition 11.7, Ce has an atomic resolution D. Again, to prove primality, we must show that any resolution B' of Ce satisfies the final hypothesis in that proposition. Now B' is non-trivially valued, so is a model of ACVF, so contains an element of any B'-definable equivalence class, as required.

The field D (obtained in either of the above cases) is a minimal resolution of e over C: for if D' is an algebraically closed subfield of D with $C \subset D'$ and $e \in \mathrm{acl}(D')$, then by primality D embeds into D' over Ce; hence as D has finite transcendence degree over C, $D = D'$.

To see uniqueness, suppose that D' is another prime resolution of Ce over C. Then D' embeds into D over Ce. By minimality of D the embedding is surjective, so D' is Ce-isomorphic to D. This gives uniqueness.

The last two assertions follow from Proposition 10.15. □

COROLLARY 11.15. *Let $E = \{e_i : i \in \omega\}$ be a countable set of imaginaries, and $C \subset K$. Then CE admits a resolution D over C such that $k(D) = k(\mathrm{acl}(CE))$ and $\Gamma(D) = \Gamma(CE)$.*

PROOF. Put $C_n = Ce_0 \ldots e_{n-1}$. Then each C_n admits a resolution D_n over C as in the theorem. The inclusions $C_n \to C_{n+1}$ yield embeddings $D_n \to D_{n+1}$. Let D be the direct limit of this system of maps.   □

COROLLARY 11.16. *Let $C \subset K$, and e a finite tuple of imaginaries. Then Ce admits a dcl-resolution D with $k(D) = k(Ce)$ and $\Gamma(D) = \Gamma(Ce)$.*

PROOF. We may assume that C is non-trivially valued, to ensure (working over C) the first assumption of Corollary 11.9. By 11.9, there exists a dcl-resolution D with $\operatorname{tp}(D/Ce)$ stationary, and embedding into any resolution. By the latter fact, and using Corollary 11.15, we have $\Gamma(D) = \Gamma(Ce)$ and $k(D) \subseteq k(\operatorname{acl}(Ce))$. But $k(D)/Ce$ is stationary, so by stable embeddedness $k(D)/k(Ce)$ is stationary, and it follows that $k(D) = k(Ce)$.   □

CHAPTER 12

MAXIMALLY COMPLETE FIELDS AND DOMINATION

We now focus on independence relations $A \underset{C}{\downarrow} B$ when all the sets A, B, C are in the field sort. Recall that a valued field is *maximally complete* if it has no proper immediate extension. In this chapter we prove that over a maximally complete base, without any assumption of stable domination, still a field is dominated by its value group and residue field (Proposition 12.11). A slightly harder but more powerful result is Theorem 12.18, that, again over a maximally complete base, the type of a field is stably dominated over its definable closure in Γ.

The following result will be used to give a criterion for orthogonality to Γ (Proposition 12.5), and for these domination results. We do not know its origin, but part (i) is Lemma 3 of [6]. It will also be important in Chapter 14. In the next proposition, if A is a subfield of K, then R_A denotes $R \cap A$ and Γ_A denotes the value group of A.

PROPOSITION 12.1. *Let $C < A$ be an extension of non-trivially valued fields, and suppose that C is maximally complete.*

(i) *Let V be a finite dimensional C-vector subspace of A. Then there is a basis $\{v_1, \ldots, v_k\}$ of V such that for any $c_1, \ldots, c_k \in C$, $|\sum_{i=1}^{k} c_i v_i| = \max\{|c_i v_i| : 1 \le i \le k\}$.*

(ii) *Assume also that $\Gamma_C = \Gamma_A$, and let $S = V \cap R_A$. Then there are generators $d_1, \ldots, d_k$ of S as a free R_C-module such that $|d_i| = 1$ for each $i = 1, \ldots, k$, and for any $f_1, \ldots, f_k \in R_C$, $|\Sigma_{i=1}^{k} f_i d_i| = \max\{|f_1|, \ldots, |f_k|\}$.*

Note that (ii) implies that $\{\mathrm{res}(d_1), \ldots, \mathrm{res}(d_k)\}$ is linearly independent over $k(C)$: for if $r_1, \ldots, r_k \in R_C$, then

$$\sum_{j=1}^{k} \mathrm{res}(r_j)\,\mathrm{res}(d_j) = 0 \iff |\sum_{j=1}^{k} r_j d_j| < 1 \iff \max\{|r_1|, \ldots, |r_k|\} < 1$$

$$\iff \mathrm{res}(r_j) = 0 \text{ for all } j.$$

We shall say that $\mathcal{B} = \{v_1, \ldots, v_k\}$ is a *separated* basis of V if (i) holds. We say in addition that it is *good* if, whenever $b, b' \in \mathcal{B}$ and $|b|\Gamma_C = |b'|\Gamma_C$, we have $|b| = |b'|$.

PROOF. (i) The separated basis is built inductively. Suppose $\{v_1, \ldots, v_\ell\}$ is a separated basis of a C-subspace U of V, and $v \in V \setminus U$.

CLAIM. There is $w \in U$ such that $|v - w| = \inf\{|v - u| : u \in U\}$.

PROOF OF CLAIM. We construct a transfinite sequence (w^ν) (ν an ordinal), where $w^\nu = \sum_{i=1}^\ell a_i^\nu v_i$, such that if $\gamma^\nu = |v - w^\nu|$ then the sequence (γ^ν) is decreasing. At stage $\nu + 1$, if $|v - w^\nu| \leq |v - x|$ for all $x \in U$, then stop. Otherwise, there is $w^{\nu+1} \in U$ with $\gamma^{\nu+1} := |v - w^{\nu+1}| < \gamma^\nu$. Now consider a limit ordinal λ. For any $\nu < \nu' < \lambda$,

$$|(v - w^\nu) - (v - w^{\nu'})| = \max\{\gamma^\nu, \gamma^{\nu'}\} = \gamma^\nu,$$

so $|\sum_{i=1}^\ell (a_i^\nu - a_i^{\nu'})v_i| = \gamma^\nu$. It follows by the inductive hypothesis that $|a_i^\nu - a_i^{\nu'}| \leq \gamma^\nu / |v_i|$ for each $i = 1, \ldots, \ell$. For each i, by choosing a pseudo-convergent subsequence of (a_i^ν) (not necessarily cofinal) and by using the maximal completeness of C, we may find some a_i^λ such that $|a_i^\lambda - a_i^\nu| \leq \gamma^\nu / |v_i|$ for all $\nu \leq \lambda$. Hence, for all $\nu < \lambda$,

$$|v - \sum_{i=1}^\ell a_i^\lambda v_i| \leq \max\{|v - \sum_{i=1}^\ell a_i^\nu v_i|, |\sum_{i=1}^\ell (a_i^\nu - a_i^\lambda)v_i|\} \leq \gamma^\nu.$$

Thus, the induction proceeds at limit stages, where we put $w^\lambda = \sum_{i=1}^\ell a_i^\lambda v_i$. As C is maximally complete, the pseudoconvergent sequence (w^λ) has a limit $w = \sum_{i=1}^\ell a_i v_i$, and this w satisfies the claim.

Given the claim, put $v_{\ell+1} := v - w$. Then $\{v_1, \ldots, v_{\ell+1}\}$ is a separated basis of the C-subspace of V which it spans: indeed, if

$$|\Sigma_{i=1}^{\ell+1} c_i v_i| < \max\{|\Sigma_{i=1}^\ell c_i v_i|, |c_{\ell+1} v_{\ell+1}|\},$$

then

$$|v_{\ell+1} + \Sigma_{i=1}^\ell c_{\ell+1}^{-1} c_i v_i| < |v_{\ell+1}|,$$

so

$$|v - (w - \Sigma_{i=1}^\ell c_{\ell+1}^{-1} c_i v_i)| < |v - w|,$$

contradicting the choice of w.

(ii) Let $\{v_1 \ldots, v_k\}$ be as in (i). Since $\Gamma_C = \Gamma_A$, for each $i = 1, \ldots, k$ there is $e_i \in C$ with $|e_i v_i| = 1$. Now put $d_i := e_i v_i$ for each i. □

LEMMA 12.2. Let $C < A$ be valued fields, and suppose that C is maximally complete. Let V be a finite or countable dimensional subspace of A (as a vector space over C). Then V has a good separated basis $\mathcal{B}$ over C.

PROOF. Suppose that $\mathcal{B}'$ is a good separated basis of the finite-dimensional subspace U of V, and let $v \in V \setminus U$. We shall show that $\mathcal{B}'$ extends to a good separated basis of $U + Cv$.

By the claim in the last proof, there is $w \in U$ such that $|v - w| = \inf\{|v - u|:$ $u \in U\}$. Put $b := v - w$. Then by the last proof, $\mathcal{B}' \cup \{b\}$ is a separated basis for the subspace (over C) which it spans.

Suppose first that for any $u \in U$, $|b| \neq |u|$. In this case, $|b| \neq \gamma|u|$ for any $\gamma \in \Gamma(C)$. Now $\mathcal{B}' \cup \{b\}$ is a separated basis of the C-space it spans (by the proof of Proposition 12.1) and still is good.

Now suppose that there is $u \in U$ with $|b| = |u|$. Put $u = \Sigma_{i=1}^m a_i b_i$, where $\mathcal{B}' = \{b_1, \ldots, b_m\}$, and $a_1, \ldots, a_m \in C$. As $\mathcal{B}'$ is separated, $|b| = |u| = |a_i b_i|$ for some i, say with $i = 1$. We claim that $|b - a_1 b_1| = |b|$. For otherwise, $|b - a_1 b_1| < |b| = |a_1 b_1|$. But by the choice of w,

$$|b - a_1 b_1| = |v - (w + a_1 b_1)| \geq |v - w| = |b|,$$

a contradiction. Now put $b_{m+1} := a_1^{-1} b$. We have $|b_{m+1}| = |b_1| = |b_{m+1} - b_1|$, the latter equality as $|a_1 b_1| = |b| = |b - a_1 b_1|$, so $|b_1| = |b a_1^{-1} - b_1| = |b_{m+1} - b_1|$. Since $\mathcal{B}' \cup \{b\}$ is a separated basis, so is $\mathcal{B}' \cup \{b_{m+1}\}$. The latter also is good, since $\mathcal{B}'$ is good and $|b_{m+1}| = |b_1|$. $\qquad\square$

REMARK 12.3. Let $C < A$ be an extension of non-trivially valued fields, with C maximally complete, and suppose that $\mathcal{B}$ is a separated basis for a subspace V of A. Then if $b_1, \ldots, b_\ell \in \mathcal{B}$ and $|b_1| = \cdots = |b_\ell|$, then $1, \mathrm{res}(b_2/b_1), \ldots, \mathrm{res}(b_\ell/b_1)$ are linearly independent over $k(C)$.

Indeed, suppose $r_1, \ldots, r_\ell \in C$ with $|r_i| \leq 1$ for each i, and

$$\Sigma_{i=1}^\ell \mathrm{res}(r_i)\,\mathrm{res}(b_i/b_1) = 0.$$

Then $|\Sigma_{i=1}^\ell r_i b_i/b_1| < 1$, so $|\Sigma_{i=1}^\ell r_i b_i| < |b_1|$. As $\mathcal{B}$ is separated, $|r_j b_j| < |b_1|$ for each j, and it follows that $|r_j| < 1$ for each j, so each $\mathrm{res}(r_j) = 0$, as required.

LEMMA 12.4. *Let $C \leq A, B$ be algebraically closed valued fields, and suppose that C is maximally complete. Assume that $\Gamma(C) = \Gamma(A)$ and that $k(A)$ and $k(B)$ are linearly disjoint over $k(C)$. Let E be the subring of K generated by $A \cup B$. Then*

(i) *Let $a_1, \ldots, a_n \in A, b_1, \ldots, b_n \in B$. Then there exist $d_1, \ldots, d_k \in A$, $b_1', \ldots, b_k' \in B$ such that in $A \otimes_C B$ we have $\sum_{i=1}^n a_i \otimes b_i = \sum_{j=1}^k d_j \otimes b_j'$ while in E we have $|e| = \max_{j=1}^k |d_j||b_j'|$.*

(ii) $\Gamma(E) = \Gamma(B)$.

(iii) *Let $a = (a_1, \ldots, a_n) \in A^n$, $b_1, \ldots, b_n \in B$, $e = \sum_{i=1}^n a_i b_i$. Let $a' = (a_1', \ldots, a_n') \models \mathrm{tp}(a/C)$. Let $e' = \sum_{i=1}^n a_i' b_i$. Then $|e'| \leq |e|$.*

(iv) *If $A' \equiv_C A$ and $k(A')$ is also linearly disjoint from $k(B)$ over $k(C)$, then $A \equiv_B A'$.*

PROOF. (i) We may suppose $|a_i| \leq 1$ for each i: for if

$$\gamma = \max\{|a_1|, \ldots, |a_n|\} > 1,$$

choose $c \in C$ with $|c| = \gamma$, and replace each a_i by $a_i c^{-1}$ and b_i by $b_i c$.

Now by Proposition 12.1 (ii) and the remark following it, there are $d_1, \ldots,$ $d_k \in A$ such that $|d_1| = \cdots = |d_k| = 1$, $\mathrm{res}(d_1), \ldots, \mathrm{res}(d_k)$ are linearly independent over $k(C)$, and $a_1, \ldots, a_n$ are in the R_C-module generated by $d_1, \ldots, d_k$. Thus, there are $c_{ij} \in R_C$ $(i = 1, \ldots, n,\ j = 1, \ldots, k)$, so that $a_i = \sum_{j=1}^{k} c_{ij} d_j$ for each $i = 1, \ldots, n$. In particular,

$$\sum_{i=1}^{n} a_i \otimes b_i = \sum_{i=1}^{n} \sum_{j=1}^{k} c_{ij} d_j \otimes b_i = \sum_{j=1}^{k} d_j \otimes \left(\sum_{i=1}^{n} c_{ij} b_i \right) = \sum_{j=1}^{k} d_j \otimes b_j',$$

where $b_j' = \sum_{i=1}^{n} c_{ij} b_i \in B$ for each j.

Let $e = \sum_{i=1}^{n} a_i b_i$; then in particular $e = \sum_{j=1}^{k} d_j b_j'$.

We may partition $\{1, \ldots, k\}$ as $I_1 \cup \cdots \cup I_m$, where for each j, $I_j := \{i: |b_i'| = \gamma_i\}$. For each $j = 1, \ldots, m$, put $e_j := \sum_{i \in I_j} d_i b_i'$. Then $e = e_1 + \cdots + e_m$. We claim that $|e_i| = \gamma_i$ for each $i = 1, \ldots, m$. So fix $i \in \{1, \ldots, m\}$, and $\ell \in I_i$. Put $f_j = b_\ell'^{-1} b_j'$ for each $j \in I_i$. Then $|f_j| = 1$ for each $j \in I_i$. As $\mathrm{res}(d_1), \ldots, \mathrm{res}(d_k)$ are linearly independent over $k(C)$, and $k(A)$ and $k(B)$ are linearly disjoint over $k(C)$, $\mathrm{res}(d_1), \ldots, \mathrm{res}(d_k)$ are linearly independent over $k(B)$. It follows that $|\sum_{j \in I_i} d_j f_j| = 1$, yielding the claim. (i) clearly follows.

(ii) Let $e \in E$. Since $\Gamma(E) \subseteq \mathrm{dcl}(\{|x|: x \in E\}$, it suffices to show $|e| \in \Gamma'(B)$. By (i) we have $e = \sum_{j=1}^{k} d_j b_j'$ with $d_j \in A, b_j' \in B$, and $|e| = \max_{j=1}^{k} |d_j||b_j'|$; so $|e| \in \Gamma(B)$.

(iii) Let d_j, b_j' be as in (i). Let $d' = (d_1', \ldots, d_k')$ be such that $\mathrm{tp}(a', d'/C) = \mathrm{tp}(a, d/C)$. Then $\sum_{i=1}^{n} a_i' b_i = \sum_{j=1}^{k} d_j' b_j'$ (since the equality already holds in the tensor product.) We have $|d_j'| = |d_j|$ using $\Gamma(A) = \Gamma(C)$. Thus $|e'| = |\sum_{j=1}^{k} d_j' b_j'| \leq \max_{j=1}^{k} |d_j||b_j'| = |e|$.

(iv) In the notation of (ii), by quantifier elimination (e.g., in $\mathcal{L}_{\mathrm{div}}$) we must show that if $(a_1', \ldots, a_n') \equiv_C (a_1, \ldots, a_n)$ and $e' := \sum_{i=1}^{n} a_i' b_i$, then $|e'| = |e|$. The proof of (i) yields this. $\square$

We deduce a criterion for orthogonality to Γ which will be used in the next chapter. It implies in particular that if C is an algebraically closed valued field with no proper immediate extensions and $a \in K^n$, then $\mathrm{tp}(a/C) \perp \Gamma$ if and only if $\Gamma(C) = \Gamma(Ca)$. Recall the notation St_C from Part I and Chapter 7. If C is a field, then $\mathrm{St}_C \subset \mathrm{dcl}(C \cup k(C))$; in general, $\mathrm{St}_C \subset \mathrm{dcl}(C \cup \mathrm{VS}_{k,C})$.

PROPOSITION 12.5. *Let $C = \mathrm{acl}_K(C)$, and let F be a maximally complete immediate extension of C, and $a \in K^n$. Then the following are equivalent.*

(i) $\mathrm{tp}(a/C) \perp \Gamma$.

(ii) $\mathrm{tp}(a/C) \vdash \mathrm{tp}(a/F)$ *and* $\Gamma(Ca) = \Gamma(C)$.

PROOF. Assume (i). Then $\Gamma(Ca) = \Gamma(C)$ by Lemma 10.2 (v). Furthermore, since the extension $C \leq F$ is immediate, $k(A) \underset{C}{\overset{g}{\downarrow}} F$, where $A = \mathrm{acl}_K(Ca)$.

As $C = \mathrm{acl}_K(C)$, $\mathrm{St}_C \subset \mathrm{dcl}(C \cup k(C))$. Thus, $\mathrm{St}_C(A) \underset{C}{\overset{}{\smile}} \mathrm{St}_C(F)$, so by Proposition 10.11, $A \underset{C}{\overset{g}{\smile}} F$ via a. Hence $\mathrm{tp}(a/C) \vdash \mathrm{tp}(a/F)$.

Conversely, suppose (ii). We first show that $\Gamma(F) = \Gamma(Fa)$. For this, it suffices to show that if $b = (b_1, \ldots, b_m) \in F^m$ then $\Gamma(Cb) = \Gamma(Cba)$. Suppose this is false, and let i be least such that $\Gamma(Cb_1 \ldots b_i) \neq \Gamma(Cb_1 \ldots b_i a)$. Then as $\mathrm{tp}(a/C) \vdash \mathrm{tp}(a/F)$, we have $\mathrm{tp}(a/Cb_1 \ldots b_{i-1}) \vdash \mathrm{tp}(a/F)$. It follows that $\mathrm{tp}(b_i/Cb_1 \ldots b_{i-1})$ implies $\mathrm{tp}(b_i/Cb_1 \ldots b_{i-1}a)$. As $\Gamma(Cb_1 \ldots b_{i-1}a) \neq \Gamma(Cb_1 \ldots b_i a)$, b_i is not in $\mathrm{acl}(Cb_1 \ldots b_{i-1}a)$, so $\mathrm{tp}(b_i/\mathrm{acl}(Cb_1 \ldots b_{i-1}))$ is not realised in $\mathrm{acl}(Cb_1 \ldots b_{i-1}a)$. Hence, by Proposition 8.22 (ii), $\Gamma(Cb_1 \ldots b_i) = \Gamma(Cb_1 \ldots b_i a)$, contradicting the choice of i.

By (ii), $a \underset{C}{\overset{g}{\smile}} F$, so to show $\mathrm{tp}(a/C) \perp \Gamma$ it suffices by Lemma 10.2 (iii) to prove $\mathrm{tp}(a/F) \perp \Gamma$. Put $A := \mathrm{acl}_K(Fa)$, and let B be an algebraically closed field extending F, with $A \underset{F}{\overset{g}{\smile}} B$. By Lemma 8.19, $k(A)$ and $k(B)$ are linearly disjoint over $k(F)$. Since $\Gamma(F) = \Gamma(A)$ (by the last paragraph), it follows from Lemma 12.4 that $\Gamma(A \cup B) = \Gamma(B)$, as required. $\square$

REMARK 12.6. Proposition 12.5 actually characterises maximally complete fields. For suppose $C = F$ is not maximally complete. Then there is a chain $(U_i : i \in I)$ of C-definable balls with no least element such that no element of C lies in $U := \bigcap(U_i : i \in I)$. Let $a \in K$ lie in U. Then $\Gamma(C) = \Gamma(Ca)$ but $\mathrm{tp}(a/C)$ is not orthogonal to Γ by Lemma 7.25.

In Chapter 9 we studied domination of a type by its stable part. Here, we examine domination of a field by its value group and residue field. For these results, we do not need to assume orthogonality to Γ, but do need the assumption that the base is a maximally complete valued field. As a consequence, we obtain the domination results (12.12, 12.18) which are the goal of this chapter. First we need several technical lemmas. Analogous to the notion of orthogonality to the value group, we have a notion of orthogonality to the residue field.

DEFINITION 12.7. Let $a = (a_1, \ldots, a_n)$. We shall say that $\mathrm{tp}(a/C)$ is orthogonal to k (written $\mathrm{tp}(a/C) \perp k$) if for any model M with $a \underset{C}{\overset{g}{\smile}} M$, we have $k(M) = k(Ma)$.

For orthogonality to k, the obvious analogues of Lemma 10.2 (i), (ii), (iii), (v) hold. Observe also that if $\mathrm{tp}(a/C) \perp k$ then $\mathrm{St}_C(a) = \mathrm{dcl}(C)$ (see the proof of Lemma 10.2 (v)).

The following is well known, but for want of a reference we give a proof.

REMARK 12.8. Let $F < L = \mathrm{acl}(F(a))$ be an extension of valued fields (where a is a finite sequence). Then

$$\mathrm{trdeg}(L/F) \geq \mathrm{trdeg}(k(L)/k(F)) + \mathrm{rk}_\mathbb{Q}(\Gamma(L)/\Gamma(F)).$$

PROOF. We may suppose that a is a singleton. Then $\operatorname{trdeg}(k(L)/k(F)) \leq 1$ and $\operatorname{rk}_{\mathbb{Q}}(\Gamma(L)/\Gamma(F)) \leq 1$. Suppose that $k(L) \neq k(F)$. Then certainly $a \notin \operatorname{acl}(F)$. We may assume that $|a| = 1$ with $\operatorname{res}(a) \notin k(F)$. Thus, a is generic in the closed ball R, and it follows from Lemma 7.25 that $\Gamma(L) = \Gamma(F)$.   □

LEMMA 12.9. *Let C, L be algebraically closed valued fields with $C \subset L$. Let $a_1, \ldots, a_m \in L$, $\gamma_i := |a_i|$ and let d_i be a code for the open ball $B_{<\gamma_i}(a_i)$ (so $d_i \in \operatorname{red}(\gamma_i R)$). Put $C' := \operatorname{acl}(C\gamma_1 \ldots \gamma_m)$, $C'' := \operatorname{acl}(Cd_1 \ldots d_m)$ and $C''' := \operatorname{acl}(Ca_1 \ldots a_m)$. Then $\operatorname{St}_{C'}(L) = \operatorname{acl}(C''k(L)) \cap \operatorname{St}_{C'}$.*

PROOF. First, consider the case $m = 0$. In this case, the conclusion of the lemma is that

$$\operatorname{St}_C(L) = \operatorname{acl}(Ck(L)) \cap \operatorname{St}_C.$$

This is clear by Lemma 7.10 (ii), since if $s \in \operatorname{acl}(C) \cap S_n$ then $\operatorname{red}(s)$ is C-definably isomorphic to k^n.

In general, we argue by induction on m. First observe that $d_i \in \operatorname{St}_{C'}(L)$ for each i. Hence $C''k(L) \subset \operatorname{St}_{C'}(L)$. As $L = \operatorname{acl}(L)$, $\operatorname{St}_{C'}(L)$ is algebraically closed in $\operatorname{St}_{C'}$, so we have the containment $\supseteq$ of the statement.

In the other direction, suppose first $C' = \operatorname{acl}(C\gamma_1 \ldots \gamma_\ell)$ for some $\ell < m$. Then by induction, we obtain

$$\operatorname{St}_{C'}(L) = \operatorname{acl}(\operatorname{acl}(Cd_1 \ldots d_\ell)k(L)) \cap \operatorname{St}_{C'} \subset \operatorname{acl}(C''k(L)).$$

Thus, we may assume that the γ_i are $\mathbb{Q}$-linearly independent in $\Gamma(L)$ over $\Gamma(C)$. It follows easily that each a_i is generic in the open ball d_i over $C''a_1 \ldots a_{i-1}$. In particular, $\operatorname{tp}(a_i/C''a_1 \ldots a_{i-1}) \perp k$: indeed, if $a_i \underset{C''a_1 \ldots a_{i-1}}{\overset{g}{\downarrow}} N$, for any model N, then as N contains a field element of d_i, $\Gamma(N) \neq \Gamma(Na_i)$, and hence $k(N) = k(Na_i)$ by Remark 12.8. From this, we obtain that $a_i \underset{C''a_1 \ldots a_{i-1}}{\overset{g}{\downarrow}} k(L)$; for example if β is a finite sequence from $k(L)$ then the above orthogonality gives $\beta \underset{C''a_1 \ldots a_{i-1}}{\overset{g}{\downarrow}} a_i$, and then Proposition 10.11 applies. Thus, $\operatorname{tp}(a_i/C''a_1 \ldots a_{i-1}k(L)) \perp k$. It follows that

$$\operatorname{tp}(a_1 \ldots a_m/C''k(L)) \perp k.$$

Hence,

$$\operatorname{St}_{C''k(L)}(C''k(L)a_1 \ldots a_m) \subseteq \operatorname{dcl}(C''k(L)),$$

so

$$\operatorname{St}_{C''k(L)}(\operatorname{acl}(C'''k(L))) = \operatorname{acl}(C''k(L)) \cap \operatorname{St}_{C''k(L)}.$$

However, applying the case $m = 0$ to the field C''', we have

$$\operatorname{St}_{C'''}(L) = \operatorname{acl}(C'''k(L)) \cap \operatorname{St}_{C'''}.$$

A fortiori, $\operatorname{St}_{C'}(L) \subset \operatorname{acl}(C'''k(L))$. So

$$\operatorname{St}_{C'}(L) \subset \operatorname{St}_{C'}(\operatorname{acl}(C'''k(L))) \subset \operatorname{St}_{C''k(L)}(\operatorname{acl}(C'''k(L))) \subset \operatorname{acl}(C''k(L)).$$

This proves the lemma. □

LEMMA 12.10. *Let C be an algebraically closed valued field. Let L, M be extension fields of C, with $\Gamma(L) \subset \mathrm{dcl}(M)$. Consider sequences $a_1, \ldots, a_r$ and $b_1, \ldots, b_s$ from L such that $|a_1|, \ldots, |a_r|$ is a $\mathbb{Q}$-basis of $\Gamma(L)$ over $\Gamma(C)$ and $\mathrm{res}(b_1), \ldots, \mathrm{res}(b_s)$ form a transcendence basis of $k(L)$ over $k(C)$ (we do not here assume that r, s are finite). Let $e_i \in M$ with $|a_i| = |e_i|$ for each i. Let $C^+ := \mathrm{dcl}(C \cup \Gamma(L))$. Then the following conditions on C, L, M are equivalent.*

(1) *For some $a_1, \ldots, a_r, b_1, \ldots, b_s, e_1, \ldots, e_r$ as above,*

$$\mathrm{res}(a_1/e_1), \ldots, \mathrm{res}(a_r/e_r), \mathrm{res}(b_1), \ldots, \mathrm{res}(b_s)$$

are algebraically independent over $k(M)$.

(2) *For all $a_1, \ldots, a_r, b_1, \ldots, b_s, e_1, \ldots, e_r$ as above,*

$$\mathrm{res}(a_1/e_1), \ldots, \mathrm{res}(a_r/e_r), \mathrm{res}(b_1), \ldots, \mathrm{res}(b_s)$$

are algebraically independent over $k(M)$.

(3) *$\mathrm{St}_{C^+}(L)$ and $\mathrm{St}_{C^+}(M)$ are independent in the stable structure St_{C^+}.*

PROOF. Let $a_1, \ldots, a_r, b_1, \ldots, b_s, e_1, \ldots, e_r$ satisfy the hypotheses of the lemma, and let d_i be the open ball of radius $|a_i|$ around a_i. By Lemma 12.9,

$$\mathrm{St}_{C^+}(L) = \mathrm{acl}(Cd_1 \ldots d_r \mathrm{res}(b_1) \ldots \mathrm{res}(b_s)) \cap \mathrm{St}_{C^+}.$$

For notational convenience we now assume r, s are finite, but the argument below, applied to subtuples, yields the lemma without this assumption. Since $\Gamma(Cb_1 \ldots b_s) = \Gamma(C)$, it follows that $d_1, \ldots, d_r, \mathrm{res}(b_1), \ldots, \mathrm{res}(b_s)$ are independent in St_{C^+}, so the Morley rank of any tuple enumerating $\mathrm{St}_{C^+}(L)$ over St_{C^+} is $r + s$. We have $\mathrm{dcl}(Md_i) = \mathrm{dcl}(M \mathrm{res}(a_i/e_i))$ for each i, as over the model M, k is in definable bijection with $\mathrm{red}(B_{\leq |a_i|}(a_i))$. Thus, (3) holds if and only if

$$\mathrm{RM}(\mathrm{tp}(\mathrm{res}(a_1/e_1), \ldots, \mathrm{res}(a_r/e_r), \mathrm{res}(b_1), \ldots, \mathrm{res}(b_s)/M)) = r + s,$$

which holds if and only if $\mathrm{res}(a_1/e_1), \ldots, \mathrm{res}(a_r/e_r), \mathrm{res}(b_1), \ldots, \mathrm{res}(b_s)$ are algebraically independent over $k(M)$. Thus (1) and (3) are equivalent. Since this equivalence is valid for any choice of $a_1, \ldots, a_r, b_1, \ldots, b_s, e_1, \ldots, e_r$ and (3) does not mention the choice, (1) and (2) are also equivalent. □

We begin with a field-theoretic version of our domination results (Proposition 12.11); and deduce a statement allowing for imaginaries (12.12). We will then deduce what turns out to be a stronger model theoretic fact, Theorem 12.18. This is be the basis of the notion of metastability. Lemma 12.10 provides an algebraic rendering of the statement of 12.18.

In the following, we say that a function f on a field induces the function h on the value group if $h(|x|) = |f(x)|$ for all x in the domain of f. Similarly, f induces h' on the residue field if $h'(\mathrm{res}(x)) = \mathrm{res}(f(x))$ for all $x \in \mathrm{dom}(f) \cap R$. If A, B, C are structures, we say that a map $f : A \to B$ (or, formally, the pair (f, B)) is unique up to conjugacy over C (subject to certain conditions) if, whenever $f_i \colon A \to B_i$ $(i = 1, 2)$ are two maps satisfying these conditions, there is an isomorphism $\ell \colon B_1 \to B_2$ over C such that $f_2 = \ell \circ f_1$. Below, we write LM (or sometimes $\langle L, M \rangle$) for the field generated by $L \cup M$.

PROPOSITION 12.11. *Let C be a maximally complete valued field, and let L, M be valued fields containing C. Let $h \colon \Gamma(L) \to \Gamma$ and $h' \colon k(L) \to k$ be embeddings, with $h(\Gamma(L)) \cap \Gamma(M) = \Gamma(C)$, and with $h'(k(L))$ and $k(M)$ linearly disjoint over $k(C)$.*

   (i) *Up to conjugacy over $M \cup h(\Gamma(L)) \cup h'(k(L))$, there is a unique pair (f, N) such that f is a valued field embedding $L \to N$ over C which induces h and h', and $\langle f(L), M \rangle = N$ (as fields).*

   (ii) *With f, N as in (i), $\Gamma(N) = \langle h(\Gamma(L)), \Gamma(M) \rangle$ (as subgroups of Γ), and $k(N) = \langle k(f(L)), k(M) \rangle$ (as fields).*

PROOF. Without loss of generality, we may assume that h and h' are the identity maps, so $\Gamma(L)$ and $\Gamma(M)$ are independent over $\Gamma(C)$ in the sense of ordered groups, and $k(L)$ is linearly disjoint from $k(M)$ over $k(C)$. To prove (i), we must show that if L' is isomorphic to L over $C \cup \Gamma(L) \cup k(L)$, then the isomorphism extends to a valued field isomorphism $LM \to L'M$ over M.

We first show that L and M are linearly disjoint over C. So suppose $u_1, \ldots, u_n \in L$ are linearly independent over C, and span a C-subspace U of L. Now U has a good separated basis $\mathcal{B} = \{b_1, \ldots, b_n\}$ by Lemma 12.2. By Remark 12.3, if $b_1, \ldots, b_\ell \in \mathcal{B}$ with $|b_1| = \cdots = |b_\ell|$ then the elements $1, \mathrm{res}(b_2/b_1), \ldots, \mathrm{res}(b_\ell/b_1)$ are linearly independent over $k(C)$. We must show that $u_1, \ldots, u_n$ are linearly independent over M, so it suffices to show that $b_1, \ldots, b_n$ are linearly independent over M. This will follow from the following claim.

CLAIM. Let $x = \Sigma_{i=1}^{n} b_i m_i$ where $m_1, \ldots, m_n \in M$. Then $|x| = \max\{|b_i||m_i| \colon 1 \le i \le n\}$.

PROOF OF CLAIM. Suppose that the claim is false. Put $\gamma := \max\{|b_i||m_i| \colon 1 \le i \le n\}$ and $J := \{i \colon |b_i m_i| = \gamma\}$. Then $|\Sigma_{i \in J} b_i m_i| < \gamma$. For any distinct $i, j \in J$ we have $|b_i/b_j| = |m_j/m_i| \in \Gamma(L) \cap \Gamma(M) = \Gamma(C)$; so $|b_i| = |b_j|$ (as $\mathcal{B}$ is good), and hence $|m_i| = |m_j|$. Fix an element (say 1) in J and let $J' := J \setminus \{1\}$. So $|b_i/b_1| = |m_i/m_1| = 1$. Now as $|\Sigma_{i \in J} b_i m_i| < \gamma$, we have $|1 + \Sigma_{i \in J'} (b_i m_i)/(b_1 m_1)| < 1$. Thus, in the residue field, $1 + \Sigma_{i \in J'} \mathrm{res}(b_i/b_1) \, \mathrm{res}(m_i/m_1) = 0$. Hence, the elements $1, \mathrm{res}(b_i/b_1)$ (for $i \in J'$) are linearly dependent over $k(M)$. Since $k(L)$ and $k(M)$ are linearly disjoint over $k(C)$, these elements are also linearly dependent over $k(C)$, contradicting the choice of the basis $\mathcal{B}$.

Now suppose that $f \colon L \to L'$ is a valued field isomorphism inducing the identity on $\Gamma(L) \cup k(L)$. Then, by the above argument, L and also L' are linearly disjoint from M over C (so independent in the sense for pure algebraically closed fields). Hence, we may extend f by the identity on M to a field isomorphism, also denoted f, from LM to $L'M$. If $x \in LM$ then $x \in UM$ for some finite dimensional C-subspace U of L with a separated basis $\{b_1, \ldots, b_n\}$, say. We may write $x = \Sigma_{i=1}^{n} m_i b_i$. Then, as in the claim $|x| = \max\{|b_i||m_i|: 1 \leq i \leq n\}$. Likewise, $|f(x)| = \max\{|f(b_i)||m_i|: 1 \leq i \leq n\}$. Since $|b_i| = |f(b_i)|$ for each i, it follows that $|x| = |f(x)|$, so f is an isomorphism of valued fields. This proof also gives that $|x| = |b||m|$ for some $b \in L$ and $m \in M$. This yields both assertions of (ii). □

In the following corollary, the conditions on residue fields and on value groups can be viewed as independence in the theories of algebraically closed fields and divisible (ordered) Abelian groups, respectively. The fields are inside $\mathcal{U}$, and types are in the sense of ACVF.

COROLLARY 12.12. *Let F be a maximally complete valued field, $F \subset A = \mathrm{dcl}(A)$. Let M be an extension of F with $k(A), k(M)$ linearly disjoint over $k(F)$, and $\Gamma(A) \cap \Gamma(M) = \Gamma(F)$. Then $\mathrm{tp}(M/F, k(A), \Gamma(A)) \vdash \mathrm{tp}(M/A)$.*

PROOF. Without loss of generality we may assume $A = \mathrm{acl}(F \cup F')$, F' finite. By Theorem 11.16 a prime dcl-resolution L of A exists and satisfies: $k(L) = k(A), \Gamma(L) = \Gamma(A)$. So the corollary for L implies, a fortiori, the same for A. Thus we may assume $A = L$ is resolved. In this case the corollary is immediate from Proposition 12.11. □

REMARK 12.13. The assumption that C is maximally complete is needed in these results, and thus the stronger ones that follow. For otherwise C has a proper extension L with $k(C) = k(L)$ and $\Gamma(C) = \Gamma(L)$. Then the hypotheses of Corollary 12.12 are vacuously true, but taking $M = A = L$, the conclusion is not.

Recall the general notion of domination of invariant types from Definition 2.2, along with the semigroup $\overline{\mathrm{Inv}}(\mathcal{U})$. We have natural embeddings of $\overline{\mathrm{Inv}}(k)$ and of $\overline{\mathrm{Inv}}(\Gamma)$ into $\overline{\mathrm{Inv}}(\mathcal{U})$. With respect to these, we have:

COROLLARY 12.14. $\overline{\mathrm{Inv}}(\mathcal{U}) \cong \overline{\mathrm{Inv}}(k) \times \overline{\mathrm{Inv}}(\Gamma)$.

PROOF. Let p be an invariant type of $\mathcal{U}$. Let C be a maximally complete algebraically closed valued field, such that $\mathrm{dcl}(C)$ is a base for p. Let $a \models p|C$, $b = k(Ca), d = \Gamma(Ca)$. Then b, d are definable functions of a, and p yields invariant types p_k, p_Γ obtained by applying these definable functions to p. It follows from Theorem 12.12 (ii) that p is domination-equivalent to $p_k \otimes p_\Gamma$. This shows that the natural homomorphism $\overline{\mathrm{Inv}}(k) \times \overline{\mathrm{Inv}}(\Gamma) \to \overline{\mathrm{Inv}}(\mathcal{U})$ is surjective. Injectivity is clear from the orthogonality of k and Γ. □

Note that $\overline{\mathrm{Inv}}(k) \cong \mathbb{N}$; the unique generator is the generic type p_{RES} of the residue field. In the next chapter we describe $\overline{\mathrm{Inv}}(\Gamma)$.

Proposition 12.11 is used in the proof of the following strengthening, leading up to Theorem 12.18.

PROPOSITION 12.15. *Let C be a maximally complete algebraically closed valued field, and let L, M be algebraically closed valued fields containing C, with L of finite transcendence degree over C. Let $g: \Gamma(L) \to \Gamma(M)$ be an embedding over C. Put $C^+ := \operatorname{dcl}(C \cup g(\Gamma(L)))$. Then, up to conjugacy over M, there is a unique pair (f, N) such that f is a valued field embedding $L \to N$ over C inducing g, $\operatorname{St}_{C^+}(f(L))$ and $\operatorname{St}_{C^+}(M)$ are independent in the stable structure St_{C^+}, and $\langle f(L), M \rangle = N$ (as fields).*

In the next lemma, and the proof of Proposition 12.15, we shall write Γ additively and use a valuation v rather than a norm, as we exploit the $\mathbb{Q}$-linear structure. If γ, δ are in the value group Γ we write $\gamma \ll \delta$ if $n\gamma < \delta$ for all $n \in \omega$. Below, when we consider places, formally ∞ is also in the range. Recall from Chapter 7 the connection between places and valuations.

LEMMA 12.16. *Let $v: L \to \Gamma$ be a valuation on a field L with corresponding place $p: L \to \operatorname{res}(L)$, let F be a subfield of $\operatorname{res}(L)$, and let $p': \operatorname{res}(L) \to F$ be a place which is the identity on F. Let $p^* = p' \circ p: L \to F$ be the composed place, with induced valuation $v^*: L \to \Gamma^*$. Suppose $a \in L$ with $p(a) \in \operatorname{res}(L) \setminus \{0\}$, and $p^*(a) = 0$.*
  (i) *If $b \in L$ with $v(b) > 0$, then $0 < v^*(a) \ll v^*(b)$,*
  (ii) *If*

$$\Delta := \{v^*(x), -v^*(x): x \in L, p(x) \notin \{\infty, 0_{\operatorname{res}(L)}\}, p^*(x) = 0\} \cup \{0_{\Gamma^*}\},$$

then Δ is a convex subgroup of Γ^* and there is an isomorphism $g: \Gamma^*/\Delta \to \Gamma$ such that $g \circ v^* = v$.

PROOF OF LEMMA 12.16. See (8.4)–(8.7) of [9], and the discussion at the end of that chapter.   □

PROOF OF PROPOSITION 12.15. The existence of f is clear, and we focus on uniqueness.

So suppose $f_i: L \to N_i$ (for $i = 1, 2$) are two maps inducing g and satisfying the hypotheses of the proposition, with v_i denoting the valuation on N_i. We will construct an isomorphism $\ell: N_1 \to N_2$ over $M \cup g(\Gamma(L))$ with $f_2 = \ell \circ f_1$. The idea is to perturb the valuations v_i to obtain valuations v_i' satisfying the additional hypothesis of Proposition 12.11 (value group independence over $\Gamma(C)$). This will yield that the v_i' are conjugate, and it will follow that the v_i are conjugate too.

Choose $a_1, \ldots, a_r \in L$ so that if $e_1 = f(a_1), \ldots, e_r = f(a_r)$, then $v(e_1), \ldots, v(e_r)$ form a $\mathbb{Q}$-basis of $g(\Gamma(L))$ over $\Gamma(C)$. Also choose $b_1, \ldots, b_s \in L$ so that $\operatorname{res}(b_1), \ldots, \operatorname{res}(b_s)$ form a transcendence basis of $k(L)$ over $k(C)$. By Lemma 12.10 applied to $f_i(L)$ and M, for $i = 1, 2$ the elements

$$\operatorname{res}(f_i(a_1)/e_1), \ldots, \operatorname{res}(f_i(a_r)/e_r), \operatorname{res}(f_i(b_1)), \ldots, \operatorname{res}(f_i(b_s))$$

are algebraically independent over $k(M)$. Let $m: k(f_1(L)) \to k(f_2(L))$ be any isomorphism over $k(C)$.

For $j = 0, \ldots, r$ and $i = 1, 2$ let

$$R_i^{(j)} := \mathrm{acl}(k(M), \mathrm{res}(f_i(a_1)/e_1),$$
$$\ldots, \mathrm{res}(f_i(a_j)/e_j), \mathrm{res}(f_i(b_1)), \ldots, \mathrm{res}(f_i(b_s)))$$

(so $R_i^{(0)} := \mathrm{acl}(k(M), \mathrm{res}(f_i(b_1)), \ldots, \mathrm{res}(f_i(b_s))))$. For each i, j choose a place $p_i^j : R_i^{(j+1)} \to R_i^{(j)}$ over $R_i^{(j)}$ (i.e., which are the identity on $R_i^{(j)}$), and a place $p_i^* : k(N_i) \to R_i^{(r)}$ over $R_i^{(r)}$. (It will eventually turn out that p_i^* is the identity and $R_i^{(r)} = k(N_i)$.) Let $p_{v_i} : N_i \to k(N_i)$ be the place corresponding to the given valuation v_i. Also let $p_{v_i'} : N_i \to R_i^{(0)}$ be the composed place $p_i^0 \circ p_i^1 \circ \cdots \circ p_i^{r-1} \circ p_i^* \circ p_{v_i}$ with corresponding valuation v_i' on N_i (determined up to value group isomorphism). Let N_i' be the valued field (N_i, v_i'), (so N_i, N_i' have the same field structure, but possibly different valuations).

Let $i \in \{1, 2\}$. Then all the p_i^j and p_i^* are the identity on $k(M)$, so p_{v_i} and $p_{v_i'}$ agree on M; so (M, v_i) and (M, v_i') are isomorphic, as a valuation is determined up to isomorphism of value groups by the corresponding place. We therefore identify (M, v_i) and (M, v_i'), identifying also the value groups.

For $i = 1, 2$ define the valued field embedding $f_i' : L \to N_i'$ by $f_i'(x) = f_i(x)$. Observe here that since $p_{v_i'}$ is the identity on $\mathrm{res}(f_i(L))$, it determines (up to isomorphism of value groups) a valuation on $f_i(L)$ which agrees with $v_i|_{f_i(L)}$. That is, for $x, y \in f_i(L)$, $v_i(x) \le v_i(y) \Leftrightarrow v_i'(x) \le v_i'(y)$. Since we have identified the value groups of (M, v_i) and (M, v_i'), we should not identify those of $(f_i(L), v_i)$ and $(f_i(L), v_i')$, but they are isomorphic. The map f_i' induces some $h_i' : \Gamma(L) \to \Gamma(N_i')$ by $h_i'(v(x)) = v_i'(x)$.

Successively applying Lemma 12.16 (with p^* replaced by $p_i^j \circ \cdots \circ p_i^{r-1} \circ p_i^* \circ p_{v_i}$), we find that

$$0 < v_i'(f_i(a_1)/e_1) \ll \cdots \ll v_i'(f_i(a_r)/e_r) \ll v_i'(b)$$

for any $b \in M$ with $v(b) > 0$. Let Δ_i be the subspace of the $\mathbb{Q}$ vector space $\Gamma(N_i')$ generated by $v_i'(f_i(a_1)/e_1), \ldots, v_i'(f_i(a_r)/e_r)$. By Lemma 12.16, Δ_i is a convex subgroup of $\Gamma(N_i')$. Since $\Gamma(L)$ and $h_i'(\Gamma(L))$ both have $\mathbb{Q}$-rank r over $\Gamma(C)$, $\Gamma(N_i') = \Delta_i \oplus \Gamma(M)$. (This yields that p_i^* is the identity, since otherwise $\Gamma(N_i')$ would have additional elements infinitesimal with respect to $\Gamma(M)$.) In particular, $h_i'(\Gamma(L)) \cap \Gamma(M) = \Gamma(C)$. Thus, there is an isomorphism $\ell^* : \Gamma(N_1') \to \Gamma(N_2')$ which is the identity on $\Gamma(M)$ and satisfies $\ell^*(v_1'(a_j)) = v_2'(a_j)$ for $j = 1, \ldots, r$. Likewise, there is a map $m^* : k(N_1') \to k(N_2')$ which is the identity on $k(M)$ and extends m.

By Proposition 12.11 there is a valued field isomorphism $\ell : N_1' \to N_2'$ which is the identity on M, extends ℓ^* and m^*, and satisfies $\ell(f_1(x)) = f_2(x)$ for all $x \in L$. We must verify that ℓ is also an isomorphism $(N_1, v_1) \to (N_2, v_2)$. However, we have $\Gamma(N_i') := \Delta_i \oplus \Gamma(M)$. Let π denote the projection map to the second coordinate. Then, by the last assertion of Lemma 12.16, for $x \in N_i$ we have $v_i(x) = \pi(v_i'(x))$. It follows that

$$\ell^*(v_1(x)) = \ell^*(\pi(v_1'(x))) = \pi(\ell^*(v_1'(x))) = \pi(v_2'(x)) = v_2(x)$$

for all $x \in N_1$, as required.   □

REMARK 12.17. Our proof gave a slightly stronger statement than Proposition 12.15. We showed that the isomorphism $\ell : N_1 \to N_2$ can be chosen to extend any given isomorphism $k(f_1(L)) \to k(f_2(L))$ which is the identity on $k(C)$.

THEOREM 12.18. (i) *Suppose that $C \leq L$ are valued fields with C maximally complete, $k(L)$ a regular extension of $k(C)$, and Γ_L/Γ_C torsion free. Let a be a sequence from $\mathcal{U}$, $a \in \mathrm{dcl}(L)$. Then $\mathrm{tp}(a/C \cup \Gamma(Ca))$ is stably dominated.*

(ii) *Let C be a maximally complete algebraically closed valued field, a a sequence from $\mathcal{U}$, and $A = \mathrm{acl}(Ca)$. Then $\mathrm{tp}(A/C, \Gamma(Ca))$ is stably dominated.*

PROOF. (i) When C is algebraically closed, this is immediate from Proposition 12.15. In general, let $\bar{C}$ be a maximally complete immediate extension of C^{alg}. The assumptions imply that

$$\mathrm{tp}(L/C) \vdash \mathrm{tp}(L/C, k(C)^{\mathrm{alg}}, \mathbb{Q} \otimes \Gamma(C)) = \mathrm{tp}(L/C, k(\bar{C}), \Gamma(\bar{C})).$$

By Corollary 12.12, $\mathrm{tp}(L/C) \vdash \mathrm{tp}(L/\bar{C})$. It follows that $\mathrm{tp}(L/C, \Gamma(L)) \vdash \mathrm{tp}(L/\bar{C}, \Gamma(L))$. But $\Gamma(L\bar{C}) = \Gamma(L)$. Since $\mathrm{tp}(L/\bar{C}, \Gamma(L\bar{C}))$ is stably dominated, so is $\mathrm{tp}(L/C, \Gamma(L))$. By Corollary 4.10 (ii), $\mathrm{tp}(a/C \cup \Gamma(Ca))$ is stably dominated.

(ii) This is a special case of (i); let $C \leq L$ be any valued field with $a \in \mathrm{dcl}(L)$.   □

REMARK 12.19. Suppose C, L are as in Theorem 12.18. Let $a_1, \ldots, a_m \in L$ be such that $|a_1|, \ldots, |a_m|$ is a $\mathbb{Q}$-basis for $\mathbb{Q} \otimes \Gamma(L)$ over $\mathbb{Q} \otimes \Gamma(C)$. Then $\mathrm{tp}(L/C \cup \Gamma(L))$ is stably dominated by $k(L) \cup \{r(a_1), \ldots, r(a_m)\}$ over $C \cup \Gamma(L)$, where $r(a_i)$ denotes the image of a_i under the reduction map $|a_i|R \to \mathrm{red}(|a_i|R)$.

PROOF. Let f enumerate $k(L) \cup \{r(a_1), \ldots, r(a_m)\}$, and let $C' = C \cup \Gamma(L)$. By Lemma 12.9, $\mathrm{St}_{C'}(L) \subseteq \mathrm{acl}(C, f)$. This, together with Theorem 12.18, gives (ii) of Proposition 3.29, and it remains to show (i), i.e., that $\mathrm{tp}(L/C', f) \vdash \mathrm{tp}(L/\mathrm{acl}(C'), f)$.

The valued field C is Henselian. Replacing L by the Henselian hull L^h will not change $\Gamma(L)$ or $k(L)$, so we may assume L is Henselian too. Thus $\mathrm{Aut}_v(L^{\mathrm{alg}}/L) = \mathrm{Aut}(L^{\mathrm{alg}}/L)$ (any field automorphism of L^{alg} over L is a valued field automorphism.) Since L is a regular extension of C, the homomorphism $\mathrm{Aut}(L^{\mathrm{alg}}/L) \to \mathrm{Aut}(C^{\mathrm{alg}}/C)$ is surjective. Equivalently, $\mathrm{tp}(C^{\mathrm{alg}}/C) \vdash \mathrm{tp}(C^{\mathrm{alg}}/L)$. But $\mathrm{dcl}(C^{\mathrm{alg}}) = \mathrm{acl}(C)$; so $\mathrm{tp}(\mathrm{acl}(C)/C) \vdash \mathrm{tp}(\mathrm{acl}(C)/L)$. Since

$C \subseteq \mathrm{dcl}(C, \Gamma(L), f) \subseteq \mathrm{dcl}(L)$, it follows a fortiori that

$$\mathrm{tp}(\mathrm{acl}(C)/C, \Gamma(L), f) \vdash \mathrm{tp}(\mathrm{acl}(C)/L).$$

Thus,

$$\mathrm{tp}(L/C, \Gamma(L), f) \vdash \mathrm{tp}(L/\mathrm{acl}(C), \Gamma(L), f)$$

But by Lemma 7.26, $\mathrm{acl}(C') = \mathrm{acl}(C \cup \Gamma(L)) = \mathrm{dcl}(\mathrm{acl}(C), \Gamma(L))$. So $\mathrm{tp}(L/C', f) \vdash \mathrm{tp}(L/\mathrm{acl}(C'), f)$ as required. $\qquad \square$

CHAPTER 13

INVARIANT TYPES

13.1. Examples of sequential independence

We give here four examples concerning issues with sequential independence. We work in the field sort throughout the chapter. In the first two examples, acl denotes field-theoretic algebraic independence (which is model-theoretic algebraic independence in ACVF in the field sort). At the end of the section, we state a result giving a setting where forking and sequential independence coincide.

EXAMPLE 13.1. As promised after Lemma 10.2, we give an example where, for $A = \mathrm{acl}(Ca)$, $\mathrm{tp}(a/C) \perp \Gamma$ even though $\mathrm{trdeg}(A/C) \neq \mathrm{trdeg}(k(A)/k(C))$. Let M be a maximally complete algebraically closed non-trivially valued field with archimedean value group. Let X be transcendental over M. There is a norm $| - |$ on $M(X)$ extending that on M, defined by $|\Sigma_{i=0}^{m} a_i X^i| = \mathrm{Max}\{|a_i|: 0 \leq i \leq m\}$. Then X is generic in R over M. Let $M\{X\}$ be the ring of convergent power series $\Sigma_{i=0}^{\infty} a_i X^i$ such that $\lim_{i \to \infty} |a_i| = 0$. Let L be its field of fractions. The norm $| - |$ extends from $M(X)$ to $M\{X\}$ by the same definition (maximum norm of coefficients), and thence to L. Choose a sequence $(n_j: j = 1, 2, \ldots)$ with $0 < n_1 < n_2 < \ldots$ and $\lim n_{j+1}/n_j = \infty$. Let $(a_i: i = 1, 2, \ldots)$ be a sequence from M with $\lim |a_i| = 0$. Put $Y := \Sigma_{i=1}^{\infty} a_i X^{n_i} \in L$. Then by the argument on p. 93 of [44], $Y \notin \mathrm{acl}(M(X))$. Now Y is a limit of a pseudoconvergent sequence $(b_i: i \in I)$ (the polynomial truncations of Y) of elements of $M(X)$, so Y is generic over $M(X)$ in a chain of $M(X)$-definable balls with no least element, namely the balls $B_{\leq |Y - b_i|}(Y)$. Thus, by Lemma 7.25, $k(\mathrm{acl}(M(X, Y))) = k(\mathrm{acl}(M(X)))$. By the definition of the norm on L, $\Gamma(L) = \Gamma(M)$, so $\Gamma(\mathrm{acl}(M(X, Y))) = \Gamma(M)$. It follows by Proposition 12.5 (with C and F of the proposition both equal to M) that $\mathrm{tp}(XY/M) \perp \Gamma$. However XY adds only 1 to the transcendence degree of the residue field, but $\mathrm{trdeg}(M(X, Y)/M) = 2$.

As a more general way to obtain such examples, let Δ be a countable divisible ordered abelian group, and k, k' be algebraically closed fields such that k' is a transcendence degree 1 field extension of k. The field $C :=$

$k((t))^\Delta$ of generalised power series of well-ordered support with exponents in Δ and coefficients in k is maximally complete; and its extension $L := k'((t))^\Delta$ has transcendence degree $2^{\aleph_0}$ over C. However, $\Gamma(L) = \Gamma(C) = \Delta$, so by Proposition 12.5, $\mathrm{tp}(C(a)/C) \perp \Gamma$ for any finite sequence a from L.

EXAMPLE 13.2. We give an example where $\mathrm{tp}(A/C) \not\perp \Gamma$, but for every singleton $a \in \mathrm{acl}(A)$, $\mathrm{tp}(a/C) \perp \Gamma$.

Let F be a trivially valued algebraically closed field of characteristic 0, let T be an indeterminate, and let the power series ring $F[[T]]$ have the usual valuation with $v(T) = 1$. As usual, we shall write $F((T))$ for the corresponding field of Laurent series. Put $C = \mathrm{acl}(F(T))$. Let $b, c \in F[[T]]$ be power series algebraically independent over C. Let F' be a trivially valued algebraically closed extension of F, and $a \in F'\backslash F$. Put $A = \mathrm{acl}(C(a, ab+c))$, and let $d \in A \setminus C$.

As $A \le F'((T))^\mathbb{Q}$, d has a Puiseux series expansion over F', so $d \in F'((T^{1/n}))$ for some n. If $d \in F((T^{1/n}))$, then $k(F(T, b, c, d)) = k(F)$, so $a \notin \mathrm{acl}(F(T, b, c, d))$. Hence, as $d \in \mathrm{acl}(F(T, a, b, c))$, we have $d \in F(T, b, c)$. This however is impossible, since by an easy argument (see e.g., p. 28 of [36]) $A \cap \mathrm{acl}(F(T, b, c)) = C$. Thus, some term $a_m T^{m/n}$ of the Puiseux series for d has $a_m \in F' \setminus F$. Choose the least such m, and put $f := \sum_{m' < m} a_{m'} T^{m'/n} \in C$. Then $g := (d - f)T^{-m/n} = a_m + e \in \mathrm{acl}(C(d))$ with $v(e) > 0$. Thus, as $a_m \notin F$, $\mathrm{res}(g) \notin k(C)$, so $\mathrm{tp}(g/C) \perp \Gamma$. As g and d are interdefinable over C, $\mathrm{tp}(d/C) \perp \Gamma$.

Suppose for a contradiction that $\mathrm{tp}(A/C) \perp \Gamma$. Then by Corollary 10.8, $\mathrm{tp}(A/C)$ is stably dominated, that is, $\mathrm{tp}(A/C)$ is dominated by $A' := k(A)$; here we use that $\mathrm{VS}_{k,C} \subseteq \mathrm{dcl}(C \cup k(C))$, as C is a field. However, as $k(C)$ is algebraic over $k(F((T)))$, we have $A' \underset{C}{\overset{s}{\perp}} F((T))$. Hence $A \underset{C}{\overset{d}{\perp}} F((T))$, so $\mathrm{tp}(A/A'C) \vdash \mathrm{tp}(A/A'Cbc)$. Thus, the formula $y = xb + c$ follows from $p(x, y) = \mathrm{tp}(a, ab+c/A'C)$, so $ab+c \in \mathrm{acl}(A'Ca)$. By quantifier elimination, as A' consists of residue field elements, $ab + c \in \mathrm{acl}(Ca)$, which is impossible.

We give two further examples concerning sequential extensions. The first takes place entirely in the value group (so in the theory of divisible ordered abelian groups), but could be encoded into ACVF. It shows that a type of the form $\mathrm{tp}(A/C)$ can have many different sequentially independent extensions, via different sequences. The second, in ACVF, shows how a type can have an invariant extension not arising through sequential independence (at least via a sequence of field elements).

EXAMPLE 13.3. We suppose that $(\Gamma, <, +, 0)$ is a large saturated divisible ordered abelian group, written additively, and with theory denoted DLO. If $C \subseteq B \subset \Gamma$ and $a \in \Gamma$, write $a \underset{C}{\overset{g}{\perp}} B$ if for any $b \in \mathrm{dcl}(B)$ with $b > a$ there is $c \in \mathrm{dcl}(C)$ with $a < c \le b$. In other words, a is chosen generically large over B within its type over C. We extend the $\underset{}{\overset{g}{\perp}}$ notation sequentially as for ACVF; it has similar properties. In particular, every type over C has an Aut(Γ/C)-invariant extension over Γ.

Note that when $F \subseteq L$ are valued fields, a, F, L contained in some $M \models$ ACVF with $C = \{v(x): x \in F\}$, $B = \{v(x): x \in L\}$, and $v(a) \notin F$, we have $a \underset{F}{\overset{g}{\bigcup}} L$ if and only if $-v(a) \underset{C}{\overset{g}{\bigcup}} B$.

Below, we write $B_n^+(\mathbb{Q})$ for the group of lower triangular $n \times n$ matrices over $\mathbb{Q}$ with strictly positive entries on the diagonal.

Let $n > 1$. We show that there are $C \subset B \subset \Gamma$ and $a = (a_1, \ldots, a_n) \in \Gamma^n$ such that if $A = \mathrm{dcl}(Ca)$, then A has infinitely many distinct sequentially independent extensions over B.

To see this, choose $C = \mathrm{dcl}(C)$ in Γ. Let $P_1, \ldots, P_n$ be cuts in C (so solution sets of complete 1-types over C), and pick $b_1 \in P_1, \ldots, b_n \in P_n$. We choose the P_i so that $0 < b_1 < \cdots < b_n$, and so that for each i, $\{x - y: y < P_i < x\} = \{x \in C: x > 0\}$. We also assume that the P_i are independent, in the sense that if $x_1 \in P_1, \ldots, x_n \in P_n$ then $\{x_1, \ldots, x_n\}$ are $\mathbb{Q}$-linearly independent over C; in particular, this choice of the x_i determines a complete n-type over C, denoted q. Put $B := \mathrm{dcl}(Cb)$ where $b = (b_1, \ldots, b_n)$. Choose $a = (a_1, \ldots, a_n)$ with $a_i \in P_i$ for each i, so that $a \underset{C}{\overset{g}{\bigcup}} B$, so a_i is chosen generically large in P_i over $Ba_1 \ldots a_{i-1}$. Let $A := \mathrm{dcl}(Ca)$.

Now for each $D \in \mathrm{GL}_n(\mathbb{Q})$, let $d := Da$ (viewing a, d as column vectors). Then $A := \mathrm{dcl}(Cd)$, and D determines an extension over B of $\mathrm{tp}(A/C)$ which is sequentially independent via d', where $d' \equiv_C d$. Let q_D be the corresponding extension over B of $\mathrm{tp}(a/C)$. Clearly q_D has an Aut(Γ/C)-invariant extension.

We claim that if D and D' lie in distinct cosets of $B_n^+(\mathbb{Q})$, then $q_D \neq q_{D'}$. Clearly $\mathrm{GL}_n(\mathbb{Q})$ acts transitively on the set of extensions over B of q of the above form, so it suffices to show that the stabiliser of $\mathrm{tp}(a/B)$ is contained in $B_n^+(\mathbb{Q})$.

Let $e_1 := a_1 - b_1, \ldots, e_n := a_n - b_n$. Then $e = (e_1, \ldots, e_n)$ is an (inverse) lexicographic basis for the ordered $\mathbb{Q}$-vector space

$$\Delta := \{0\} \cup \{x \in \mathrm{dcl}(Ba): 0 < |x| < c \text{ for all } c \in C^{>0}\}.$$

Now the flag

$$(0, \mathbb{Q}e_1, \mathbb{Q}e_1 + \mathbb{Q}e_2, \ldots, \mathbb{Q}e_1 + \cdots + \mathbb{Q}e_n)$$

is uniquely determined by the ordered vector space structure on Δ; the kth element is the unique convex subspace of Δ of dimension k.

We must show that if $D \in \mathrm{GL}_n(\mathbb{Q})$, $d = Da$, and $a \underset{C}{\overset{g}{\bigcup}} B$ and $d \underset{C}{\overset{g}{\bigcup}} B$ both hold, then $D \in B_n^+(\mathbb{Q})$. So suppose $d \underset{C}{\overset{g}{\bigcup}} B$. If $b' := Db$ with $b' := (b_1', \ldots, b_n')$, then d_i is chosen in the cut over C containing b_i'. Thus, arguing as above, there will be $y_1, \ldots y_n \in B$ such that

$$d_1 - y_1 \in \mathbb{Q}^+ e_1, \ldots, d_n - y_n \in \mathbb{Q}e_1 + \cdots + \mathbb{Q}e_{n-1} + \mathbb{Q}^+ e_n.$$

It is easily checked that if this holds then $D \in B_n^+(\mathbb{Q})$.

In the rest of the chapter, the field valuation will be written additively. In the next lemma, and the example which follows, we will use the completion $\bar{F}$ of a valued field F with value group Δ. For convenience, we assume that Δ is countable and Archimedean, so in particular has countable cofinality. The completion is a standard construction; it can be obtained as the field of limits of Cauchy sequences from F, where (a_n) is Cauchy if $v(a_{n+1} - a_n)$ is increasing and cofinal in Δ. The field $\bar{F}$ is characterized by the facts that $v(\bar{F}) = v(F)$, every Cauchy sequence in $\bar{F}$ converges, and every element of $\bar{F}$ is the limit of a sequence from F. It is easy to see that $\bar{F}$ is algebraically closed if F is.

We look at sequentially generic extensions (in the sense of ACVF) of types over F of tuples from $\bar{F}$; this is interesting for its own sake, and will also be used below. We assume F is algebraically closed, $v(F) = \Delta \neq (0)$, and $\mathrm{trdeg}_F(\bar{F}) > 1$. For any field L, L^a denotes its algebraic closure. Examples with $F \neq \bar{F}$ (both algebraically closed) include Krasner's $F = (\mathbb{Q}_p)^a$, with completion $\mathbb{C}_p$; or $F = F_0(t)^a$, F_0 a countable field; note that $\bar{F}$ is uncountable.

Suppose $F \subset F'$ where $F', \bar{F} \subset M$ (a sufficiently saturated algebraically closed valued field), and F' admits a valued field embedding φ over F into $\bar{F}$. In this case, φ is unique. Working inside M, if $d \in F'$ then $\varphi(d)$ is the unique element of $\bar{F}$ with $v(d - \varphi(d)) > \Delta$.

Let $\rho(d) = v(d - \varphi(d))$, and $\bar{\rho}(d) = \rho(d) + \Delta \in \Gamma(M)/\Delta$. Also, let $P\bar{\rho}(x) = p^{\mathbb{Z}}\bar{\rho}(x)$, where $p = 1$ if $\mathrm{char}(F) = 0$, and $p = \mathrm{char}(F)$ otherwise. (So $P\bar{\rho}(x)$ gives less information than $\bar{\rho}(x)$ but a bit more than the $\mathbb{Q}$-space $\mathbb{Q}\bar{\rho}(x)$.)

LEMMA 13.4. *Assume n is finite, $F' = F(d_1, \ldots, d_n)^a$ embeds into $\bar{F}$ over F via φ as above, and $\Delta < \rho(d_1) \ll \rho(d_2) \ll \cdots \ll \rho(d_n)$. Then $P\bar{\rho}$ takes just n non-zero values on F', namely the $P\bar{\rho}(d_i)$.*

Before proving this, we mention a fact which is presumably well-known.

LEMMA 13.5. *Let $h = h(x_0, \ldots, x_n)$ be an irreducible polynomial over a field L. Let $h(a) = 0$, where $\mathrm{trdeg}(L(a)/L) = n$. Let $G(y_0, \ldots, y_n) := h(a_0 + y_0, \ldots, a_n + y_n)$ be the Taylor expansion about a. Fix i between $0, \ldots, n$. Let $G_i(y_i) = G_i(0, \ldots, 0, y_i, 0, \ldots, 0)$, and write $G_i(y_i) = g_i(y_i^{q_i})$ where q_i is 1 in characteristic 0, and in characteristic p, q_i is a power of p chosen so that g_i is a polynomial with some monomial of degree not divisible by p. Then g_i has nonzero term of degree 1, and zero constant term.*

PROOF OF LEMMA 13.5. First, g has zero constant term as $h(a) = 0$. We may suppose $i = 0$. Then by the genericity of the zero a, $h(x, a_1, \ldots, a_n)$ is irreducible over $L(a_1, \ldots, a_n)$, and G_0 is the expansion of h about a_0. So we may assume $n = 0$. Write $h(x) = H(x^q)$, where q is a power of p and H has a monomial of degree not divisible by p (in characteristic 0, $q = 1$). Then the derivative H' is not identically 0. Now $\deg_x H'(x^q) < \deg_x H(x^q) = \deg h$, so

$H'(x^q)$ cannot vanish at a root of h, by the irreducibility of h. So $H'(a_0^q) \neq 0$. However, this is the linear coefficient of the expansion of h about a_0. $\qquad \square$

PROOF OF LEMMA 13.4. Let $d_0 \in F' \setminus F$; we wish to compute $p(d_0)$. The elements $d_0, \ldots, d_n$ are algebraically dependent over F. If some proper subset $d_0, d_{i_1}, \ldots, d_{i_k}$ is algebraically dependent, then we can use induction. So assume they are not. Let $d = (d_0, \ldots, d_n)$. We have $h(d) = 0$ for some irreducible $h \in F[X_0, \ldots, X_n]$; and $g(d) \neq 0$ for all nonzero g of smaller total degree.

Let $d_i' = \varphi(d_i), d' = (d_0', \ldots, d_n'), e_i = d_i' - d_i$. Then $h(d') = 0$.

Using multi-index notation, expand $h(d_0 + y_0, \ldots, d_n + y_n) = \sum h_\nu(d) y^\nu$, with $h_\nu \in F[X_0, \ldots, X_n]$. So, since $h(d') = 0$,

$$\sum h_\nu(d) e^\nu = 0.$$

In particular, the lowest two values $v(h_\nu(d) e^\nu)$ must be equal.

Now as F' embeds into $\bar{F}$ over F, $v(F') = v(F) = \Delta$. So $v(h_\nu(d)) \in \Delta$. Note that $h_\nu(d) \neq 0$ whenever $h_\nu \neq 0$, since h_ν has smaller degree than h.

Consider the various monomials $h_\nu(d) y^\nu$. Since $d_0, d_2, \ldots, d_n$ are independent, y_1 occurs in h; otherwise, $h(d)$ does not involve d_1. By Lemma 13.5, we may write $h(d_0 + y_0, \ldots, d_n + y_n) = g(y_1^{q_1}) + f(d, y)$, where: g is a polynomial in 1 variable (with coefficients polynomial in the d_i), $q_1 = 1$ in characteristic 0, and in characteristic p, q_1 is a power of p, g has zero constant term and non-zero linear term, and each monomial in f involves some y_i other than y_1. Thus, if ν_1 is the multi-index corresponding to $y^{\nu_1} = y_1^{q_1}$, then $h_{\nu_1} \neq 0$, so $h_{\nu_1}(d) \neq 0$.

Any ν involving some y_j, $j > 1$ has: $\Delta < v(e^\nu) - v(e^{\nu_1})$. Thus in looking for the summands of smallest valuation, such ν can be ignored. The same holds for any ν involving y_1 to a higher power than q_1. Note that by definition of φ we have $v(e_0) = v(d_0 - d_0') = v(d_0 - \varphi(d_0)) > \Delta$. So $v(e_0 e_1) - v(e_1) > \Delta$ and so terms involving $y_0 y_1$ must be of higher valuation than e_1 also.

This leaves indices y^ν of the form y_0^m only. Let ν_0 be the least such; $y^{\nu_0} = y_0^{q_0}$, q_0 least possible. (Again, $q_0 = 1$ in characteristic 0, and is a power of p in characteristic p.) As noted above, $v(e_0) > \Delta$. Then terms $h_\nu(d) e^\nu$ where $y^\nu = y_0^k$, $k > m_0$, again cannot have least value among the summands, since they have value greater than $h_{\nu_0}(d) e^{\nu_0}$. We are left with only two candidates for summands of least value; so they must have equal value.

$$v(h_{\nu_1}(d) e^{\nu_1}) = v(h_{\nu_0}(d) e^{\nu_0}).$$

So $v(e_0) \in (q_1/q_0) v(e_1) + \Delta$. Thus $p(d_0) = (q_1/q_0) v(e_1) + \Delta$, and so $P\bar{p}(d_0) = P\bar{p}(d_1)$. $\qquad \square$

Working in the theory DLO, let Γ be a large saturated model of DLO, Δ a small subset of Γ with no greatest element, and $B \subset \Gamma$. Let $q^\Delta(u)$ denote the $\mathrm{Aut}(\Gamma/\Delta)$-invariant 1-type over Γ containing the formulas $u > \gamma$ for all $\gamma \in \Delta$, and $u < \gamma$ for all $\gamma \in \Gamma$ with $\Delta < \gamma$. For $B \subset \Gamma$, let q_B^Δ denote the restriction $q^\Delta | B$.

Let $F, M \models ACVF$ with $\bar{F} \subset M$. Let $\Delta := \Gamma(F)$ and $c \in \bar{F} \setminus F$. For any $L \supset F$, consider the generic extension to L (in the $\underset{\smile}{\;}^g$-sense) of $p_c(x) = \text{tp}(c/F)$. Denote this by $p_c | L$. Then, if $c \in L$, $p_c | L$ contains all formulas of the DLO-type $q_L^\Delta(v(x - c))$. Indeed, this determines $p_c | L$. Note that $p_c | \mathcal{U}$ is $\text{Aut}(\mathcal{U}/F)$-invariant.

We extend this notation for 2-types. If $c = (c_1, c_2)$ with $c_i \in \bar{F} \setminus F$, and $L \supseteq F$, let $p_c(x_1, x_2) | L$ be the sequentially generic extension over L of $\text{tp}(c_2, c_1/F)$ (so we take the x_2 variable first). Then if $F \models ACVF$ and $L \supseteq \bar{F}$, $p_c | L$ is implied by: $q_L^\Delta(v(x_2 - c_2)), q_{Lx_2}^\Delta(v(x_1 - c_1))$.

EXAMPLE 13.6. There is $F \models ACVF$, and $A = \text{acl}(A) \supset F$, such that $\text{tp}(A/F)$ has an $\text{Aut}(\mathcal{U}/F)$-invariant extension p over $\mathcal{U}$, such that if A' realises p then there is no acl-generating sequence d of field elements from A' with $d \underset{F}{\overset{g}{\smile}} \mathcal{U}$.

To construct this, we choose F as above, let $\Delta := \Gamma(F)$, and pick $c_1, c_2 \in \bar{F}$, algebraically independent over F.

For any $B \supseteq F$, let $q_B := q_B^\Delta$. Then let $s_B(u, v)$ be the 2-type over B determined by specifying $q_B(u)$ and $q_{Bu}(v)$. Let D be any invertible matrix over $\mathbb{Q}$ which is not lower-triangular, chosen so that the rows are positive elements of $\mathbb{Q}^2$ in the lexicographic order. Then let $r_B(u, v)$ be the 2-type of DLO given as

$$(\exists x, y)(s_B(x, y) \wedge (u, v) = D(x, y)).$$

Then r_B is a complete 2-type over B. Finally, $t_B^{c_1, c_2}(x_1, x_2)$ holds if we have $r_B(v(x_1 - c_1), v(x_2 - c_2))$.

Now $t_B^{c_1, c_2}$ is a complete 2-type in the field sort over any $B \supseteq F c_1 c_2$. Furthermore, by its definition, $t_{\mathcal{U}}^{c_1, c_2}$ is $\text{Aut}(\mathcal{U}/F c_1 c_2)$-invariant.

CLAIM. $t_{\mathcal{U}}^{c_1, c_2}$ is $\text{Aut}(\mathcal{U}/F)$-invariant.

PROOF OF CLAIM. Let $\sigma \in \text{Aut}(\mathcal{U}/F)$, and put $(d_1, d_2) := \sigma((c_1, c_2))$. We must show $t_{\mathcal{U}}^{c_1, c_2} = t_{\mathcal{U}}^{d_1, d_2}$. Thus, we must show $t_{\mathcal{U}}^{c_1, c_2}$ includes $r_{\mathcal{U}}(v(x_1 - d_1), v(x_2 - d_2))$. Suppose N is a model containing $F c_1 c_2 d_1 d_2$, and (a_1, a_2) realises $t_N^{c_1, c_2}$. By the definition of s, r, and t, we have $v(a_1 - c_1) < v(b)$ and $v(a_2 - c_2) < v(b)$ for any $b \in N$ with $v(b) > \Delta$. If $\gamma \in \Delta$ then by the condition on the c_i, there is for each i some $c_i' \in F$ with $v(c_i' - c_i) > \gamma$, and hence also $v(c_i' - d_i) = v(c_i' - \sigma(c_i)) > \gamma$. Thus, $v(c_i - d_i) > \gamma$ for all $\gamma \in \Delta$. It follows that $v(a_i - c_i) < v(d_i - c_i)$ for each i. Thus, $v(a_i - d_i) = v(a_i - c_i)$, so $r_{\mathcal{U}}(v(a_1 - d_1), v(a_2 - d_2))$ holds.

Now suppose $a = (a_1, a_2) \models t_{\mathcal{U}}^{c_1, c_2}$, and $A = (Fa)^{\text{alg}}$, $F' := (F c_1 c_2)^{\text{alg}}$, $B := (F'a)^{\text{alg}}$. Let $p := \text{tp}(A/\mathcal{U})$. Suppose that $A \underset{F}{\overset{g}{\smile}} \bar{F}$ via a sequence $d' = (d_1', \ldots, d_m')$ of field elements. By considering transcendence degree,

$m = 2$. Now A embeds into $\bar{F}$ over F. Hence, by the remarks before the example, $d' \models p_d$ for some $d = (d_1, d_2) \in \bar{F}^2$.

We adopt the notation $\varphi, \rho, \bar{\rho}, P\bar{\rho}$ from Lemma 13.4. Let $N := \bar{F}(d')^{\text{alg}}$ and $F'' := F(d')^{\text{alg}}$. Clearly, $\Gamma(N) = \Delta + \mathbb{Q}\rho(d'_1) + \mathbb{Q}\rho(d'_2)$, with $\Delta \ll \rho(d'_1) \ll \rho(d'_2)$. In Lemma 13.4, we studied all possible sequentially generic types of this form over F; and noticed that for such a type p_d, $P\bar{\rho}$ takes just two values on $F'' \setminus F$. One of these values lies in the unique proper nonzero convex subspace S of $\Gamma(N)/\Delta$. However, if $(a_1, a_2) \models t_{\mathcal{U}}^{c_1, c_2} | \bar{F}$, then, by the choice of D, $P\bar{\rho}(a_1) \neq P\bar{\rho}(a_2)$ and neither of these two values corresponds to S. □

13.2. Invariant types, dividing and sequential independence

We proceed towards a theorem asserting that invariant extensions, non-forking, non-dividing and sequential independence are all essentially the same, over sufficiently good base structures. Our previous results on stable domination reduce this rather quickly to the case of Γ. Most of the effort in this section is thus concentrated on Γ.

Let A, B be divisible linearly ordered Abelian groups. By a *cut* in A we mean a partition $A = L^- \cup L^+$, such that $a < b$ whenever $a \in L^-, b \in L^+$. We will sometimes describe a cut using a downward-closed subset $L' \subseteq A$ alone, having in mind $(L', L \setminus L')$.

We let $A^{\geq 0} = \{a \in A : a \geq 0\}$. Let $|h| = \pm h \geq 0$. If $A \leq B$ and $b \in B$, $b > 0$, the *cut of* b over A is $\text{ct}_A(b) = \{a \in A : 0 \leq a \leq b\}$. If (X^-, X^+) is any cut in A, the *corresponding convex subgroup* is defined to be $H(X) := \{h \in A : h + X^- = X^-, h + X^+ = X^+\}$. We write $\text{Hct}_A(b) = H(\text{ct}_A(b))$.

An embedding $A \subset B$ of divisible ordered Abelian groups is called an *i-extension* of A if there exists no $b \in B$ with $\text{ct}_A(b)$ closed under addition. Equivalently, the map $H \mapsto (H \cap A)$ is a bijection between the convex subgroups of A and of B; the inverse map then takes a convex subgroup H' of A to the convex hull of H' in B. This shows transitive of i-extensions.

A divisible ordered Abelian group A is *i-complete* if it has no proper i-extensions. We will show that any divisible ordered abelian group has an i-complete i-extension, unique up to isomorphism.

The algebraically closed valued field C will be called *vi-complete* if it is maximally complete, and the value group is i-complete.

Let $C \leq A, B$ be subgroups of a divisible ordered Abelian group. We say that A is *i-free* from B over C if there are no $b_1 \leq a \leq b_2$, $b_1, b_2 \in B$, $a \in A \setminus C$, with $\text{ct}_C(b_1) = \text{ct}_C(b_2)$. This implies in particular that $A \cap B = C$. If $A = C(a)$, we will also say that $\text{tp}(a/B)$ is an *i-free* extension of $\text{tp}(a/C)$.

Below, we use the notion of sequential independence in Γ from Definition 8.2 and Remark 7.18 (iii), rather than that of Example 13.3. In particular, $a \downarrow_C^g B$ if for any $b \in \mathrm{dcl}(B)$ with $b \le a$, there is $c \in \mathrm{dcl}(C)$ with $b \le c \le a$.

THEOREM 13.7. *Let $C \le B$ be algebraically closed valued fields, and A be a field which is finitely generated over C. Suppose that C is vi-complete. Then the following are equivalent.*

(i) $\mathrm{tp}(A/B)$ *does not fork over C.*

(ii) $\mathrm{tp}(A/B)$ *does not divide over C.*

(iii) $k(A)$ *and $k(B)$ are free over $k(C)$, and $\Gamma(A)$ is i-free from $\Gamma(B)$ over $\Gamma(C)$.*

(iv) *A is sequentially independent from B over C via some sequence of generators.*

(v) $\mathrm{tp}(A/B)$ *has an $\mathrm{Aut}(\mathcal{U}/C)$-invariant extension over $\mathcal{U}$.*

The number of extensions of $\mathrm{tp}(A/C)$ to B satisfying these equivalent properties is at most $2^{\dim_{\mathbb{Q}}(\Gamma(A)/\Gamma(C))}$.

Note that by Example 13.6, the implication (v) $\Rightarrow$ (iv) fails without some assumption on C.

REMARK 13.8. ACVF is an NIP theory, that is it does not have the *independence property* of [49]; this is an easy consequence of quantifier elimination. By a recent observation of Shelah [51] and also Adler, for any complete NIP theory T with sufficiently saturated model $\mathcal{U}$ and some elementary submodel M, if $p \in S(\mathcal{U})$ then p is $\mathrm{Aut}(\mathcal{U}/M)$-invariant if and only if it does not fork over M. Thus, the equivalence (i) $\Leftrightarrow$ (v) above does not require the vi-assumption on M.

From here to the proof of the theorem, all lemmas (13.9 to 13.18) refer purely to the category of divisible ordered Abelian groups. The material is in part similar to [50].

LEMMA 13.9. *Every divisible ordered Abelian group A has an i-complete i-extension.*

PROOF. Take a transfinite chain A_α of proper i-extensions of A; it is clear that the union of a chain of i-extensions is again one. Thus it suffices to show that the chain stops at some ordinal, i.e., to bound the size of any i-extension B of A. We show the bound $|B| \le 2^{2^{|A|}}$, which can be proved rapidly and suffices for our purposes, though a more careful analysis should give $2^{|A|}$. For $b_1 \ne b_2 \in B$, let $c(\{b_1, b_2\}) = \mathrm{ct}_A(|b_1 - b_2|)$. Suppose $|B| > 2^{2^{|A|}}$. By the Erdös–Rado theorem ([25] Theorem 69 and Ex. 69.1), there exist $y < y' < y'' \in B$ and $C \subseteq A$, such that $\mathrm{ct}_A(y' - y) = \mathrm{ct}_A(y'' - y) = \mathrm{ct}_A(y'' - y') = C$. It follows that C is closed under addition; this contradicts the definition of an i-extension. $\square$

LEMMA 13.10. *Let $X = (X^-, X^+)$ be a cut in $A^{\ge 0}$, $H = H(X)$. Let $X' = (X'^-, X'^+)$ be the image of X in A/H. Then the following are equivalent:*

1. *There exists an i-extension B of A and $b \in B$ with $\mathrm{ct}_A(b) = X$.*
2. *X'^{-} has no maximal element, and X'^{+} has no minimal element.*

PROOF. If X'^{-} has a maximal element $a + H$, then the coset $a + H$ is cofinal in X^{-}; so if $\mathrm{ct}_A(b) = X^{-}$ then $\mathrm{ct}_A(b - a) = H \cup A^{<0}$. If A/H has a minimal element $d + H$ above X'^{-} then $\mathrm{ct}_A(d - a) = H^{\geq 0} \cup A^{<0}$. Both cases contradict the definition of an i- extension.

In the converse direction, form $B = A(b) = \{a + \alpha b : a \in A, \alpha \in \mathbb{Q}\}$, with $a + \alpha b < a' + \alpha'b$ iff $(a - a')/(\alpha' - \alpha) \in X$.

To show that B is an i-extension, let $J \subset B$ be a convex subgroup, $J' = J \cap A$. Let $\alpha b + a \in J$. We must find $a' \in J \cap A$, $\alpha b + a < a'$. Multiplying all by $|1/\alpha|$, we may assume $\alpha = \pm 1$.

Replacing X by the dual cut $(-X^{+}, -X^{-})$ corresponds to replacing b by $-b$. Hence it suffices to prove the claim for positive α, so we may assume $\alpha = 1$.

Note that $a + X^{-}$ is downward closed, and is not the cut of a convex subgroup (else X^{-} is the cut of $H - a$, but then the image of X^{-} in A/H has a maximal element.) Thus there exists $c \in a + X^{-}$ with $2c > a + X^{-}$. Let $a' = 2c$. Then $X^{-} < a' - a$ while $(1/2)(a' - 2a) \in X^{-}$. So $(1/2)(a' - 2a) < b < a' - a$ and $b + a < a' < 2(b + a)$. As $a + b \in J$, $2(a + b) \in J$, so $a' \in J$ as required. $\square$

COROLLARY 13.11. *Let A be i-complete, X a downward-closed subset of A. Let $H = H(X)$. Then there exists $a \in A$ such that*

$$X = \{x : (\exists h \in H)(x \leq h + a)\}$$

or

$$X = \{x : (\forall h \in H)(x < h + a)\}$$

PROOF. By Lemma 13.10 , X^{-} has a maximal element $a + H$, or else there exists a minimal element $a + H$ above X^{-}. $\square$

Cuts of the form $\{x : (\exists h \in H)(x \leq h)\}$ or $\{x : (\forall h \in H)(x < h)\}$ will be called cuts of convex subgroups. The converse of Corollary 13.11 is also easily seen to be true: if every cut of A is a translate of a cut of a convex subgroup, then A is i-complete.

The following uniqueness statement will not be used but is included as background.

PROPOSITION 13.12. *Up to A-isomorphism, A has a unique i-complete i-extension.*

PROOF. Let B_1, B_2 be two i-complete i-extensions of A. Let $f : B_1 \to B_2$ be an isomorphism with maximal domain $A' \supseteq A$. If $A' = B_1$, $fA' = B_2$ we are done. Otherwise say $A' \neq B_1$, $c_1 \in B_1 \setminus A'$. We may assume notationally that $f = \mathrm{id}_{A'}$. Let X be the lower cut of c_1 over A', and let H be the corresponding convex group. By Lemma 13.10, as the cut is realized in the i-extension B_1, the image of X in A'/H has no least upper bound. (In particular, X has no maximal element.) Now $H_2 := H_{B_2}$, the convex hull of H in B_2, is the group belonging to the convex closure X_2 of X in B_2. For if $h \in H$, then $h + X \subseteq X$, and so $h + X_2 \subseteq X_2$. Thus $H_2 \subseteq H(X_2)$. Conversely if $h \in H(X_2) \cap A'$, $h + X_2 \subseteq X_2$, so $h + X \subseteq X_2 \cap A = X$, hence $h \in H$. Thus $H(X_2) \cap A' = H$. By the contrapositive of Lemma 13.10 and by the i-completeness of B_2, the image of X_2 in B_2/H_2 does have a least upper bound $c + H_2$. Observe that this coset of H_2 cannot be represented in A'; for if $(c + H_2) \cap A' \neq \emptyset$, then it is a least upper bound for the image of X in A'/H; but such a bound does not exist.

Now for $x \in X_2$, $c - x \geq h$ for some $h \in H_2$, while for $x > X_2$, $c - x \leq h$ for some $h \in H_2$. As H is cofinal in H_2 (in both directions), we may take $h \in H$. Now if $x \in X$, then $c - x \geq h$ for some $h \in H$, so $c \geq x + h$. If $c \leq x$, then $x + h \leq c \leq x$ so $x + H_2 = c + H_2$, contradicting the fact that $(c + H_2) \cap A' = \emptyset$. Thus $X < c$. If $c \geq x$, $x \in A'$, $x > X$, then $c - x \leq h$ for some $h \in H$. So $x \leq c \leq x + h$ and again $c + H = x + H$, contradiction. Thus the cut of c over A' is precisely X. Hence f may be extended to an isomorphism $A'(c_1) \rightarrow A'(c)$, contradicting the maximality of f.

The following definition is phrased in terms of a universal domain $\mathcal{U}$ for the theory of divisible ordered Abelian groups; C, A', A, B are substructures. Compare 12.12.

DEFINITION 13.13. Let $C \leq A' \leq A$ be divisible ordered Abelian groups. A is *i-dominated* by A' over C if for any $B \supset C$ with A' i-free from B over C we have: $\mathrm{tp}(B/A') \vdash \mathrm{tp}(B/A)$.

Let $C \leq A$. Given a cut P of C, let $P_A = \{a \in A : \mathrm{ct}_C(a) = P\}$.

LEMMA 13.14. *Let* $C \subset A' \subset A$ *be divisible ordered Abelian groups. Assume: for every cut* P *of* C, $P_{A'}$ *is (left) cofinal in* P_A. *Then* A *is i-dominated by* A' *over* C.

Note that using the reflection $x \mapsto -x$, one sees that the right-sided condition is the same.

PROOF. Given $B \supset C$ with A' i-free from B over C, the isomorphism type over A, B of the group $B + A$ is determined since $B \cap A' = C$, and the ordering is also determined as follows:

$$b + a_1 \leq b' + a_2 \iff b - b' < a_2 - a_1$$
$$\iff (\exists a' \in A')(b - b' \leq a' \leq a_2 - a_1) \qquad \square$$

If H is a convex subgroup of A, let $q_H = q_H | A$ denote the type over A of an element c with $\mathrm{ct}_c(A) = H \cup A^{<0}$. If $b_1 \models q_H$, then H remains a convex subgroup in $A(b_1) = A + \mathbb{Q}b_1$. Let $b_2 \models q_H | (A(b_1))$, and let $q_H^2 | A = \mathrm{tp}((b_2, b_1)/A)$. Proceed in this way to define q_H^n. Thus $(c_1, \ldots, c_n) \models q_H^n$ iff

$$H < c_n \ll c_{n-1} \ll \cdots \ll c_1 < (A^{\geq 0} \setminus H)$$

If $A \leq B$, $q_H(B)$ denotes the set of realizations of q_H in B.

LEMMA 13.15. *Let $H_1, \ldots, H_k$ be distinct convex subgroups of A. Let B be an extension of A generated by tuples b_i realizing $q_{H_i}^{m_i}$. Then the isomorphism type of B over A is determined. In particular, the coordinates of the elements b_i are $\mathbb{Q}$-linearly independent over A. Thus if B is finitely generated over A, then the cuts of only finitely many convex subgroups of A are realized in B.*

PROOF. We have to show that $\cup_{i=1}^{k} q_{H_i}^{m_i}(x_i)$ is a complete type. Say $H_1 \subset \cdots \subset H_k$, and use induction on k. Let $A' = A(b_2, \ldots, b_k)$. By induction, the A-isomorphism type of A' is determined. It is easy to check that H_1 is a convex subgroup of A', and $q_{H_1}^{m_1}$ generates a complete type over A', namely $q_{H_1} | A'$. $\square$

LEMMA 13.16. *Let A be i-complete, $A \subset B$, B/A finitely generated. Then there exist distinct convex subgroups $H_1, \ldots, H_k$ of A, and $m_1, \ldots, m_k \in \mathbb{N}$, such that $q_{H_i}^{m_i}$ is realized in B by some tuple b_i, $b = (b_1, \ldots, b_k)$, and B is i-dominated by $A(b)$ over A.*

PROOF. If $q_{H_1}^{m_1} \times \cdots \times q_{H_k}^{m_k}$ is realized in B then by Lemma 13.15 we have $\sum_{i=1}^{k} m_i \leq \mathrm{rk}_{\mathbb{Q}}(B/A)$. Thus a maximal b as in the statement of the lemma exists. We claim that B is i-dominated by $A(b)$ over A.

Let P be a cut of A. By Lemma 13.14 it suffices to show that $P_{A(b)}$ is two-sided cofinal in P_B.

Since A is i-complete, possibly after inversion, P can be translated by an element of A, to a cut of a convex group H; we may assume $P = \{x: (\exists h \in H)(x < h)\}$ (Lemma 13.11). Suppose for contradiction that $a \in P_B$, but all elements of $P_{A(b)}$ are below a (or all above a). Say $H = H_i$ (add a term with $m_i = 0$ if necessary). Then (a, b_i) (respectively (b_i, a)) realizes $q_{H_i}^{m_i+1}$. This contradicts the maximality in the choice of b. $\square$

LEMMA 13.17. *Let $C \leq B$. Let H be a convex subgroup of C.*
(1) *q_H has at most two i-free extensions to B.*
(2) *q_H^m has at most $m + 1$ i-free extensions to B.*
(3) *Let $H_1 < H_2 < \cdots < H_n$ be convex subgroups of C, $q = q_{H_1}^{m_1} \otimes \cdots \otimes q_{H_n}^{m_n}$. The number of i-free extensions of q to B is at most $\Pi_{i=1}^{n}(1 + m_i) \leq 2^{\sum_{i=1}^{n} m_i}$.*

PROOF. (1) Let $a \models q_H$, with $C(a)$ i-free from B over C. By i-freeness, we have either $r_1(a): H < a < q_H(B)$ or $r_2(a): q_H(B) < a < C^{\geq 0} \setminus H$. Either of these two possibilities determines a complete type over B. (Note that $q_H(B)$ is closed under multiplication by $\mathbb{Q}^{>0}$.)

(2) Let $(a_1, \ldots, a_m) \models q_H^m$. By (1) for some l we have $a_i \models r_1$ for $i \leq l$ and $a_i \models r_2$ for $i > l$. It is easy to see that $r_1(x_1) \cup \cdots \cup r_1(x_l) \cup x_l \ll b$ (for $b \in q_H(B)$)$\cup q_H^l(x_1, \ldots, x_l) \cup r_2(x_{l+1}) \cup \cdots \cup r_2(x_m) \cup q_H^{m-l}(x_{l+1}, \ldots, x_m)$ generates a complete type. This gives $m + 1$ possibilities.

(3) For each i, there are $\leq m_i + 1$ i-free extensions $q_{H_i}^{m_i}$; and it is easy to see that for any system r_i of choices of such extensions, $r_1(x_1) \cup \cdots \cup r_n(x_n)$ is a complete type. $\qquad\square$

LEMMA 13.18. *Let A be generated over C by realizations of $q_{H_i}^{m_i}$ for some finite set $H_1 \subset \cdots \subset H_n$ of distinct convex subgroups C. Let $C \leq B$. If A is i-free from B over C, then A is g-free from B over C, via some sequence of generators, by definition.*

PROOF. This reduces to the case $n = 1$ and $m_1 = 1$. Let $a \in A$, $a \models q_H | C$. If $a < b$ for some $b \in B$ with $b \models q_H$, then by the i-freeness assumption, $a < b$ for all such b. In this case, $\mathrm{tp}(a/B)$ is g-free over C (in the sense of Remark 7.18 (iii), where the valuation is replaced by a norm).

Otherwise, $a > b$ for all $b \in B$ with $b \models q_H$. Then $-a$ is g-free over C, and we use $-a$ as the generator. $\qquad\square$

PROOF OF THEOREM 13.7. (i) implies (ii) by definition.

(ii) implies (iii): If $\mathrm{tp}(A/B)$ does not divide over C, then the same is true of $\mathrm{tp}(k(A)/k(C))$ and $\mathrm{tp}(\Gamma(A)/\Gamma(C))$. Hence we are reduced to showing the implication inside k and Γ. For k it is well-known that non-dividing agrees with linear disjointness (over an algebraically closed field.) As for Γ, suppose $\Gamma(A)$ is not i-free from $\Gamma(B)$ over $\Gamma(C)$. Then there exist $a \in \Gamma(A) \setminus \Gamma(C)$, $b_0, b_0' \in \Gamma(B)$ such that $b_0 \leq a \leq b_0'$ and b, b' lie in the same cut over $\Gamma(C)$. Extend (b_0, b_0') to an indiscernible sequence $((b_i, b_i'): i < \omega)$ over C with $b_0 < b_0' < b_1 < b_1' \ldots$. Clearly the formulas $b_i < x < b_i'$ are pairwise inconsistent, showing that $\mathrm{tp}(\Gamma(A)/\Gamma(B))$ divides over $\Gamma(C)$.

(iv) implies (v) is Corollary 8.15.

(v) implies (i) by definition of forking (see Remark 2.9).

It remains to prove that (iii) implies (iv). Observe that if $C \leq A' \leq A$ and $\mathrm{tp}(B/A') \vdash \mathrm{tp}(B/A)$, and if A' is sequentially independent from B via some sequence of generators a', then A is sequentially independent from B via any (unary) sequence of generators whatever extending a'. Moreover any extension of $\mathrm{tp}(A'/C)$ to B extends uniquely to an extension of $\mathrm{tp}(A/C)$ to B, so the number of extensions with property (iii) is the same for A' and for A. Hence in this case the theorem is reduced from A to A'.

Since $\mathrm{tp}(B/C, k(A), \Gamma(A)) \vdash \mathrm{tp}(B/A)$, we may assume $A = k(A) \cup \Gamma(A)$, $B = k(B) \cup \Gamma(B)$. As k, Γ are orthogonal, we can consider them separately. Clearly $k(A)$ is free from $k(B)$ over $k(C)$, and we are reduced to Γ. Similarly, by Lemma 13.16, we may assume A is generated over C by realizations of $q_{H_i}^{m_i}$ for some distinct convex subgroups C_i of C. This is given by Lemma 13.18.

The final assertion follows from Lemma 13.17. $\qquad\square$

REMARK 13.19. Forking is not symmetric even over a vi-complete C; this is because i-freeness is not in general symmetric. However one can find, given $\text{tp}(A/C)$ and $\text{tp}(B/C)$, a 2-type that is non-forking in both directions.

For an $\text{Aut}(\mathcal{U}/C)$-invariant convex subgroup H we define an invariant type q_H as follows: if $B \leq \Gamma$, $c \models q_H|B$ if $H \cap B < c < B^{\geq 0} \setminus H$. Given a small subset A of Γ, let

$$H^- = H^-(A) = \{x \in \Gamma : (\forall a \in A)(\forall n \in \mathbb{N})(n|x| \leq a)\}$$
$$H^+(A) = \{x \in \Gamma : (\exists a \in A)(\exists n \in \mathbb{N})(|x| \leq na)\}$$

Also define $^-q_A = q_{H^-}$, $^+q_A = q_{H^+}$.

If $A = \emptyset$ we set $^-q_A = q_\Gamma$, while ^+q_A is not defined.

In this notation, we have $^-q_A =^- q_{A'}$ if A, A' are left-cofinal in each other, and similarly for $+$. Let $\mathcal{Q} = \{^-q_A, ^+q_A : A \text{ small.}\}$.

Note that if B is a small subgroup, a convex subgroup H of B corresponds to two invariant types, ^+q_H and ^-q_J where $J = \{b \in B : b > H\}$. Both restrict to $q_H|B$. When $B \leq B'$, and $c \models q_H|B$, we have: $B(c)$ is i-free from B' over B iff c realizes one of these types over B'. Let r be one of the types ^+q_H and ^-q_J. The notion '$c \models r|B'$' is thus finer than i-freeness, and correspondingly domination is easier than i-domination. Indeed it is easy to see that r^2 is domination-equivalent to r.

COROLLARY 13.20. $\overline{\text{Inv}}(\Gamma)$ is the free Abelian semigroup generated by $\mathcal{Q}$, subject to the identity: $x^2 = x$.

PROOF. Let q be an invariant type, with base A; we may take A to be i-complete. Note that if $A \leq B$ then $q|B$ is an i-free extension of $q|A$. By Lemma 13.16, $q|B$ follows from $q|A$ together with $q'|B$, where q' is a certain type of the form $\otimes_i q_{H_i}^{m_i}$. Thus we are reduced to considering the case: $q = q_H$, H a convex subgroup of A. In this case by Lemma 13.17, q has at most two invariant extensions to $\mathcal{U}$, namely ^-q_H and $^+q_{A^{\geq 0}\setminus H}$. Thus $\overline{\text{Inv}}(\Gamma)$ is generated by the said types. The freeness is easy and left to the reader. $\square$

This should be combined with Corollary 12.14.

CHAPTER 14

A MAXIMUM MODULUS PRINCIPLE

In this chapter, we look at the relationship between sequential independence and a form of independence, called *modulus independence*, which is clearly symmetric. If A and B are fields, and $C \cap K \leq A$ with $\Gamma(C) = \Gamma(A)$, we show that sequential independence over C implies modulus independence. This can be understood as a maximum modulus principle: if a polynomial function takes a certain norm on a realisation of a type, then it takes at least that norm on a generic point of the type (over parameters defining the function). Using the results of Chapter 10, we conclude by showing that for fields, if there is some orthogonality to the value group, then sequential independence and modulus independence are equivalent.

DEFINITION 14.1. Suppose $A \subset K$. Recall that we write x_a for a tuple of variables corresponding to the elements a of A. Define

$$\text{tp}^<(A/C) := \{|f(x_a)| < \gamma : f(X) \in \mathbb{Z}[X] \text{ and } \text{tp}(A/C) \vdash |f(x_a)| < \gamma\},$$

$$\text{tp}^\leq(A/C) := \{|f(x_a)| \leq \gamma : f(X) \in \mathbb{Z}[X] \text{ and } \text{tp}(A/C) \vdash |f(x_a)| \leq \gamma\},$$

and

$$\text{tp}^+(A/C) := \text{tp}^<(A/C) \cup \text{tp}^\leq(A/C).$$

Because of the assumption that $C \cap K \leq A$, nothing is changed if we allow the polynomials f in $\text{tp}^+(A/C)$ to have coefficients in $C \cap K$ rather than $\mathbb{Z}$. In the above, γ ranges over Γ. We have however:

LEMMA 14.2. *Let $P(X)$ be a consistent partial type over C in field variables, and let $f \in (C \cap K)[X]$ be a polynomial. Let $\gamma \in \Gamma$. Then:*

$$P \vdash |f(X)| \leq \gamma \text{ iff for some } \gamma \geq \gamma_0 \in \Gamma(C), \ P \vdash |f(X)| \leq \gamma_0$$

$$P \vdash |f(X)| < \gamma \text{ iff either } \gamma \in \Gamma(C)$$

$$\text{or for some } \gamma > \gamma_0 \in \Gamma(C), \ P \vdash |f(X)| \leq \gamma_0.$$

PROOF. If $P \vdash |f(X)| \leq \gamma$ then for some $\psi \in P$ we have $\mathcal{U} \models (\forall x)(\psi(x) \to |f(X)| \leq \gamma)$. Let $I = \{\gamma' \in \Gamma : \mathcal{U} \models (\forall x)(\psi(x) \to |f(X)| \leq \gamma')\}$. Then I is C-definable, nonempty and closed upwards. We cannot have $I = \Gamma$, since if $c \models P$ and $\gamma < |f(c)|$ then $\gamma \notin I$. So $I = [\gamma_0, \infty)$ or $I = (\gamma_0, \infty)$, with $\gamma_0 \in \Gamma(C)$. It suffices to show that $P \models |f(X)| \leq \gamma_0$. Otherwise, let $a \models P, |f(a)| > \gamma_0$. Choose γ' with $\gamma_0 < \gamma' < |f(a)|$. Then $\gamma' \notin I$ according to the definition of I, but $\gamma' \in (\gamma_0, \infty)$, a contradiction.   □

Observe that $\mathrm{tp}^<$ determines $\mathrm{tp}^{\leq}$. For suppose $\mathrm{tp}^<(a/C) = \mathrm{tp}^<(a'/C)$, and $\mathrm{tp}(a/C) \vdash |f(x)| \leq \gamma$. Then for all $\delta > \gamma$, $\mathrm{tp}^<(a/C) \vdash |f(x)| < \delta$, so $\mathrm{tp}^<(a'/C) \vdash |f(x)| < \delta$, and hence $\mathrm{tp}(a'/C) \vdash |f(x)| \leq \gamma$.

Nonetheless, we use both strict and weak inequalities; this permits using the above lemma to see that the type of $\mathrm{tp}^+(A/C)$ consisting of inequalities with parameters $\gamma \in \Gamma(C)$ implies the full type.

DEFINITION 14.3. If A and B are subsets of K such that

$$\mathrm{tp}(A/C) \cup \mathrm{tp}(B/C) \vdash \mathrm{tp}^+(AB/C)$$

we will say that A and B are *modulus-independent* over C and write $A \underset{C}{\overset{m}{\bigcup}} B$.

The condition $A \underset{C}{\overset{m}{\bigcup}} B$ says that if h is any elementary map over C, then $Ah(B)$ satisfies $\mathrm{tp}^+(AB/C)$. Observe that $\overset{m}{\bigcup}$ is symmetric, and is independent of any choice of generating sequence for A over C, so does not coincide with $\overset{g}{\bigcup}$ (by Examples 8.4 and 8.5). The main theorem of the chapter shows that, under certain hypotheses, sequential independence implies modulus independence. This gives a consequence of $\overset{g}{\bigcup}$-independence which is symmetric between left and right.

THEOREM 14.4. *Let* $C = \mathrm{acl}((C \cap K) \cup H)$ *where* $H \subset S = \bigcup_{m>0} S_m$, *and let* A, B *be valued fields, with* $C \subseteq \mathrm{dcl}(A) \cap \mathrm{dcl}(B)$ *and* $\Gamma(C) = \Gamma(A)$. *Suppose* $A \underset{C}{\overset{g}{\bigcup}} B$, *via a sequence* $a = (a_1, \ldots, a_n)$ *of field elements. Then* $A \underset{C}{\overset{m}{\bigcup}} B$.

The proof of Theorem 14.4 requires a sequence of lemmas; some will also be useful elsewhere. The case where $\mathrm{tp}(A/C) \perp \Gamma$ requires only a few of these; see the proof of Theorem 14.12.

LEMMA 14.5. *Let* A, B, C *be valued fields, with* $C \leq A \cap B$, *and* C *algebraically closed and maximally complete. Suppose* $\Gamma(C) = \Gamma(A)$ *and* $k(A)$ *and* $k(B)$ *are linearly disjoint over* $k(C)$. *Then,*

(i) *let* $g \in B[X]$ *be a polynomial over* B, $a \in A^n$, *and* $\gamma \in \Gamma(B)$, *and suppose* $|g(a)| \leq \gamma$ *(respectively* $|g(a)| < \gamma$*). Then* $\mathrm{tp}(a/C) \vdash |g(x)| \leq \gamma$ *(respectively* $|g(x)| < \gamma$*).*

(ii) $A \underset{C}{\overset{m}{\bigcup}} B$.

PROOF. Note that by Lemma 12.4, $\Gamma(B) = \Gamma(AB)$. We argue just with $<$ as the argument is identical .

(i) This is Lemma 12.4 (iii).

(ii) Let $f(X, Y) \in C[X, Y]$. We need to show that if a, b are finite tuples from A, B respectively, and $\mathrm{tp}(AB/C) \vdash |f(x_a, y_b)| < \gamma$, then

$$\mathrm{tp}(A/C) \cup \mathrm{tp}(B/C) \vdash |f(x_a, y_b)| < \gamma,$$

and similarly with $\leq$ replacing $<$. By Lemma 14.2, we may assume $\gamma \in \mathrm{dcl}(C)$. Letting $g(x) = f(x, e)$, we need to prove:

$$\mathrm{tp}(a/C) \vdash |g(x)| \leq \gamma.$$

This follows from (i) (where the same is shown more generally for $\gamma \in \Gamma(B)$).

$\square$

COROLLARY 14.6. *Let* $C = \mathrm{acl}(C) \leq \mathcal{U}$, *and let* B *be a valued field with* $C \leq \mathrm{dcl}(B)$. *Let* $g \in B[X]$. *Assume* p *is a stably dominated* $\mathrm{Aut}(\mathcal{U}/C)$-*invariant type in the field sort,* $a \models p|B$, *and* $\gamma = |g(a)|$. *Then* $p|C \vdash |g(x)| \leq \gamma$.

PROOF. Note $\gamma \in \Gamma(B)$, since p is orthogonal to Γ.

Let M be any algebraically closed, maximally complete valued field with $C \subseteq \mathrm{dcl}(M)$, and such that $\mathrm{St}_C(M) \underset{C}{\perp} \mathrm{St}_C(B)$. Let $a \models p|M(b)$, where b enumerates B. By the stable domination of p, the hypotheses of Lemma 14.5 apply to $M(a), M(b), M$. Hence $p|M \vdash |g(x)| \leq \gamma$. Hence $p|Ce \vdash |g(x)| \leq \gamma$ for some finite tuple e from M. Let I be a large index set, and let $(e_i : i \in I)$ be an indiscernible sequence over B with $e_1 = e$ and $(\mathrm{St}_C(e_i) : i \in I)$ a Morley sequence over B. Since $\mathrm{St}_C(e) \underset{C}{\perp} \mathrm{St}_C(B)$, this is also a Morley sequence over C. Let $a' \models p|C$. By Lemma 2.11, $\mathrm{St}_C(a') \underset{C}{\perp} \mathrm{St}_C(e_i)$ for some i; so by stable domination, $a' \models p|Ce_i$. Since $\mathrm{tp}(e/B) = \mathrm{tp}(e_i/B)$, we have $p|Ce_i \vdash |g(x)| \leq \gamma$. So $|g(a')| \leq \gamma$. Since $a' \models p|C$ was arbitrary, $p|C \vdash |g(x)| \leq \gamma$.

$\square$

LEMMA 14.7. *Let* A, B, C, C' *be fields with* $C \leq A \cap B \cap C'$. *Put* $A' := \mathrm{acl}_K(AC')$ *and* $B' := \mathrm{acl}_K(BC')$. *Assume*

(i) $\mathrm{tp}(A/C) \vdash \mathrm{tp}(A/C')$, *and*

(ii) $A' \underset{C'}{\overset{m}{\perp}} B'$.

Then $A \underset{C}{\overset{m}{\perp}} B$.

PROOF. By (ii), $\mathrm{tp}(A'/C') \cup \mathrm{tp}(B'/C') \vdash \mathrm{tp}^+(AB/C')$, so $\mathrm{tp}(A/C') \cup \mathrm{tp}(B/C') \vdash \mathrm{tp}^+(AB/C')$. By (i), it follows that

$$\mathrm{tp}(A/C) \cup \mathrm{tp}(B/C') \vdash \mathrm{tp}^+(AB/C'),$$

so

$$\mathrm{tp}(A/C) \cup \mathrm{tp}(B/C') \vdash \mathrm{tp}^+(AB/C).$$

It follows that $\mathrm{tp}(A/C) \cup \mathrm{tp}(B/C) \vdash \mathrm{tp}^+(AB/C)$. For suppose $a \in A, b \in B$, $\gamma \in \Gamma(C)$, and $|f(a,b)| < \gamma$ (where f is over $\mathbb{Z}$). Let σ be an automorphism (or elementary map) on K over C. We must show that $|f(a, \sigma(b))| < \gamma$ (and similarly with $<$ replaced by $\leq$). But clearly $\mathrm{tp}(\sigma(A)/C) \cup \mathrm{tp}(\sigma(B)/\sigma(C')) \vdash \mathrm{tp}^+(\sigma(AB)/C)$, and $|f(\sigma(a), \sigma(b))| \leq \gamma$. Hence, as $a \equiv_C \sigma(a)$, we have $|f(a, \sigma(b))| < \gamma$.

$\square$

LEMMA 14.8. *Let* A, B, C *be valued fields, with* C *algebraically closed,* $C \leq A, B$. *Suppose* $\mathrm{tp}(A/C) \perp \Gamma$ *and* $A \underset{C}{\overset{g}{\perp}} B$. *Then* $A \underset{C}{\overset{m}{\perp}} B$.

PROOF. Let C' be a maximally complete immediate extension of C, chosen so that $A \underset{C}{\overset{g}{\smile}} BC'$. Then by Proposition 12.5, $\mathrm{tp}(A/C) \vdash \mathrm{tp}(A/C')$, and $\Gamma(AC') = \Gamma(C')$. Put $A' := \mathrm{acl}(AC')$ and $B' := \mathrm{acl}(BC')$. Then $A' \underset{C'}{\overset{g}{\smile}} B'$, so by Lemmas 8.19 and 14.5, $\mathrm{tp}(A'/C') \cup \mathrm{tp}(B'/C') \vdash \mathrm{tp}^+(A'B'/C')$. The lemma now follows from Lemma 14.7. $\qquad\square$

LEMMA 14.9. *Suppose $C = \mathrm{acl}(C) \cap K$, $a_1 \in K$, and $C \subseteq B \subset K$. Assume also $\Gamma(C) = \Gamma(Ca_1)$ and $a_1 \underset{C}{\overset{g}{\smile}} B$. Let $a_2 \in K$ be interalgebraic with a_1 over C. Then $a_2 \underset{C}{\overset{g}{\smile}} B$.*

PROOF. Since $\Gamma(C) = \Gamma(Ca_1)$, $\mathrm{tp}(a_1/C)$ is not the generic type of an open ball. By Lemma 10.9, the result is immediate if a_1 is generic in a closed ball over C, so we may suppose that a_1 is generic in the intersection of a sequence of C-definable balls $(U_i : i \in I)$ with no least element, and that the U_i are closed. Then $\mathrm{tp}(a_1/B)$ is determined by a collection of formulas

$$\{x \in U_i : i \in I\} \cup \{x \notin V_j : j \in J\},$$

where the V_j are B-definable balls (possibly of radius zero) all lying in $\bigcap(U_i : i \in I)$. Suppose that $a_2 \underset{C}{\overset{g}{\not\smile}} B$. Then there is a B-definable ball V containing no C-definable balls, such that $a_2 \in V$. Let $\varphi(x, y)$ be a formula over C satisfied by $a_1 a_2$ and implying that x and y are interalgebraic over C. Then

$$\{x \in U_i : i \in I\} \cup \{x \notin V_j : j \in J\} \vdash \exists y (\varphi(x, y) \wedge y \in V).$$

Hence, by compactness, there is $i_0 \in I$ such that

$$\{x \in U_{i_0}\} \cup \{x \notin V_j : j \in J\} \vdash \exists y (\varphi(x, y) \wedge y \in V). \qquad (*)$$

Suppose that a_1' is generic over B in U_{i_0}. Then by $(*)$, there is a_2' interalgebraic over C with a_1' with $a_2' \in V$. Then $a_2' \underset{C}{\overset{g}{\not\smile}} B$. Since $\mathrm{tp}(a_1'/C) \perp \Gamma$, this contradicts Lemma 10.9. $\qquad\square$

PROOF OF THEOREM 14.4. Put $H = \{s_i : i < \lambda\}$. We may suppose $A = \mathrm{acl}_K(A)$. For each i, let $A_i := \mathrm{acl}_K(Ca_1 \ldots a_{i-1})$, and let U_i be the intersection (in K) of the A_i-definable balls containing a_i. Let δ be the number of i such that U_i is not a single ball, that is, such that there is no smallest A_i-definable ball containing a_i. We argue by induction on δ, over all possible A, B, C satisfying the hypotheses of the theorem. The strategy is to reduce δ by replacing a chain of balls with no least element by a closed ball chosen generically inside it.

First, we reduce to the case when $H = \emptyset$, that is, $C = \mathrm{acl}(C \cap K)$. So for fixed δ, we assume the result holds when $H = \emptyset$. Let $A^* \equiv_C A$ and $B^* \equiv_C B$. We must show $\mathrm{tp}^+(AB/C) = \mathrm{tp}^+(A^*B^*/C)$. Let $(e_i : i < \lambda)$ be a generic closed resolution of $(s_i : i \in \lambda)$ over $CABA^*B^*$ (so for each i, e_i is a generic resolution of s_i over $CABA^*B^* \cup \{e_j : j < i\}$). Put $E := \bigcup(e_i : i < \lambda)$ (treating each e_i as a subset of K). Then $e \underset{C}{\overset{g}{\smile}} ABA^*B^*$ for each finite $e \subset E$, and $\mathrm{tp}(e/C) \perp \Gamma$ (by Remark 10.4 (iii)), so $ABA^*B^* \underset{C}{\overset{g}{\smile}} e$, and hence

$ABA^*B^* \mathrel{\underset{C}{\overset{g}{\textstyle\bigcup}}} E$. Put $C' := \mathrm{acl}(CE)$, $A' := \mathrm{acl}_K(AE)$, and $B' := \mathrm{acl}_K(BE)$. Now as $H \subset \mathrm{dcl}(B)$, $e \mathrel{\underset{B}{\overset{g}{\textstyle\bigcup}}} A$ (e as above), and $\mathrm{tp}(e/B) \perp \Gamma$. Hence, by Proposition 10.11, $A \mathrel{\underset{B}{\overset{g}{\textstyle\bigcup}}} Be$, so $A \mathrel{\underset{B}{\overset{g}{\textstyle\bigcup}}} B'$, both via a. Now as $A \mathrel{\underset{C}{\overset{g}{\textstyle\bigcup}}} B$, we have $A \mathrel{\underset{C}{\overset{g}{\textstyle\bigcup}}} B'$, so $A \mathrel{\underset{C'}{\overset{g}{\textstyle\bigcup}}} B'$, and so $A' \mathrel{\underset{C'}{\overset{g}{\textstyle\bigcup}}} B'$, all via a. Also, as the resolution is generic, $\Gamma(A') = \Gamma(A) = \Gamma(C)$. We claim that the invariant δ for (A', B', C') is no greater than for (A, B, C). For suppose it is greater. Then for some i, there is a chain of $\mathrm{acl}(C'a_1 \ldots a_{i-1})$-definable closed balls which contain a_i but are not $\mathrm{acl}(Ca_1 \ldots a_{i-1})$-definable. Let s be (a code for) one such closed ball. Then, as $\Gamma(A') = \Gamma(C)$, $\gamma := \mathrm{rad}(s) \in \Gamma(C)$, and for any generic closed resolution E' of H over CA, $s = B_{\leq\gamma}(a_i) \in \mathrm{acl}(Ca_1 \ldots a_{i-1}E')$. Hence, as $H \subset C$, $s \in \mathrm{acl}(Ca_1 \ldots a_{i-1})$, proving the claim. Since the elements of H are definable over E, $C' = \mathrm{acl}(C' \cap K)$, and it follows (by the case when $H = \emptyset$) that $\mathrm{tp}(A'/C') \cup \mathrm{tp}(B'/C') \vdash \mathrm{tp}^+(A'B'/C')$. Hence, $\mathrm{tp}(A/C') \cup \mathrm{tp}(B/C') \vdash \mathrm{tp}^+(AB/C')$. Also, as $AA^* \mathrel{\underset{C}{\overset{g}{\textstyle\bigcup}}} E$ and $BB^* \mathrel{\underset{C}{\overset{g}{\textstyle\bigcup}}} E$, it follows by Corollary 8.13 that $\mathrm{tp}(A/C') = \mathrm{tp}(A^*/C')$ and $\mathrm{tp}(B/C') = \mathrm{tp}(B^*/C')$. Hence, A^*B^* satisfies $\mathrm{tp}^+(AB/C')$ and in particular $\mathrm{tp}^+(AB/C)$, as required.

Thus, we assume $C = \mathrm{acl}(C \cap K)$. In particular, each A_i is resolved. We may suppose there is an immediate extension C^* of C such that $\mathrm{tp}(A/C)$ does not imply a complete type over C^*. For otherwise, by Proposition 12.5 (ii) $\Rightarrow$(i), $\mathrm{tp}(a/C) \perp \Gamma$, so Lemma 14.8 is applicable. Choose C^* so that $a \mathrel{\underset{C}{\overset{g}{\textstyle\bigcup}}} C^*$. We may now find $C \leq C' \leq C_1 \leq C^*$ so that $\mathrm{tp}(A/C)$ implies a complete type over C', but not over C_1, and so that $\mathrm{trdeg}(C_1/C') = 1$.

Next, we reduce to the case when $C = C'$. To justify this, put $A' := \mathrm{acl}_K(AC')$, and $B' := \mathrm{acl}_K(BC')$. Then $\Gamma(C') = \Gamma(A')$. Indeed, if not, then for some i, a_i is chosen generically over $\mathrm{acl}(C'a_1 \ldots a_{i-1})$ in a $(C'a_1 \ldots a_{i-1})$-∞-definable 1-torsor U which is not closed and contains a proper subtorsor V also definable over $\mathrm{acl}(C'a_1 \ldots a_{i-1})$. As $a_i \mathrel{\underset{Ca_1\ldots a_{i-1}}{\textstyle\bigcup}} C'$, U is ∞-definable over $\mathrm{acl}(Ca_1 \ldots a_{i-1})$. Since $\Gamma(C) = \Gamma(Ca)$ and a_i is generic in U over $Ca_1 \ldots a_{i-1}$, there is no $\mathrm{acl}(Ca_1 \ldots a_{i-1})$-definable proper subtorsor of U. Hence, if $a_i' \in V$, then $a_i' \equiv_{Ca_1 \ldots a_{i-1}} a_i$ but $a_i' \not\equiv_{C'a_1 \ldots a_{i-1}} a_i$. This contradicts the assumption that $\mathrm{tp}(a/C) \vdash \mathrm{tp}(a/C')$. Thus, we can apply the argument given below, with (C', A', B') in place of (C, A, B) to obtain: $A' \mathrel{\underset{C'}{\overset{m}{\textstyle\bigcup}}} B'$. Then Lemma 14.7 yields $A \mathrel{\underset{C}{\overset{m}{\textstyle\bigcup}}} B$.

It follows that we may assume $C = C'$, so $\mathrm{trdeg}(C_1/C) = 1$. We have $a \mathrel{\underset{C}{\overset{g}{\textstyle\bigcup}}} C_1$. Pick $c_1 \in C_1 \setminus C$, and let V be the intersection of all C-definable balls containing c_1. As C^* is an immediate extension of C, $k(C_1) = k(C)$ and so V is not closed. Also $\Gamma(C_1) = \Gamma(C)$, so V contains no C-definable proper subtorsors.

We now let h be a C-elementary map, and must show $Ah(B)$ satisfies $\mathrm{tp}^+(AB/C)$. Let i be largest such that $\mathrm{tp}(A_i/C) \vdash \mathrm{tp}(A_i/C_1)$. Then $\mathrm{tp}(A_{i+1}/C) \not\vdash \mathrm{tp}(A_{i+1}/C_1)$ so by Proposition 8.22 (i) there is an embedding of C_1 into A_{i+1} over C. Hence, there is $a_i' \in A_{i+1}$ with $a_i' \in V$. Furthermore, as $\mathrm{tp}(A_i/C) \vdash \mathrm{tp}(A_i/C_1)$, $\mathrm{tp}(c_1/C) \vdash \mathrm{tp}(c_1/A_i)$. Hence, $x \in V$ determines a complete type over A_i; for otherwise, different choices of c_1 in V would have different types over A_i whilst having the same type over C. In particular, $a_i' \notin A_i$, so a_i and a_i' are inter-algebraic over A_i. Since C is resolved and C_1 is an immediate extension of C, there is no smallest C-definable ball containing V; for if V was an open ball, then c_1 would be generic over C in an open ball, so $\Gamma(C_1)$ would properly contain $\Gamma(C)$. Hence, using Lemma 7.25 twice, $k(A_i) = k(A_i a_i') = k(A_i a_i)$, so a_i is not generic in a closed ball over A_i. As $\Gamma(C) = \Gamma(A)$, a_i is also not generic in an open ball over A_i, so $\mathrm{tp}(a_i/A_i)$ is generic in a chain of balls with no least element. In particular, $\delta > 0$. Since we assume the result holds for smaller values of δ, our goal is to decrease i by converting the choice of a_i over A_i to the choice of a generic in a closed ball.

Choose $\alpha \in \Gamma$ generic over $CABh(B)$ below the cut in $\Gamma(\mathcal{U})$ made by $\mathrm{rad}(V)$. Let $u := B_{\leq\alpha}(a_i')$. Also, let s be a code for the lattice in S_2 interdefinable over $\emptyset$ with u (as in Remark 7.4). Put $\hat{C} := \mathrm{acl}(C\alpha s) = \mathrm{acl}(Cs)$ (as $\alpha \in \mathrm{dcl}(s)$). By Proposition 8.8 (ii), there is an elementary map f over C_2 such that $A \underset{\hat{C}}{\overset{g}{\downarrow}} f(B)$ via a. Put $a' := (a_1, \ldots, a_{i-1}, a_i', a_{i+1}, \ldots, a_n)$.

CLAIM. (i) $A \underset{C}{\overset{g}{\downarrow}} f(B)$ via a'.

  (ii) $A \underset{\hat{C}}{\overset{g}{\downarrow}} f(B)$ via a'.

  (iii) $A \underset{C}{\overset{g}{\downarrow}} B$ via a'.

PROOF OF CLAIM. (i) First, for $j < i$, note that $\mathrm{tp}(a_j/A_j)$ implies a complete type over Cc_1 and hence over Ca_i' (as $a_i' \equiv_{A_j} c_1$). Hence, as $\Gamma(C) = \Gamma(A)$, $\mathrm{tp}(a_j/A_j)$ implies a complete type over $A_j a_i' \alpha$, and hence over $\mathrm{acl}(A_j s\alpha)$. Thus, $(a_1, \ldots, a_{i-1}) \underset{C}{\overset{g}{\downarrow}} \hat{C}$, so as $A \underset{\hat{C}}{\overset{g}{\downarrow}} f(B)$, we have

$$(a_1, \ldots, a_{i-1}) \underset{C}{\overset{g}{\downarrow}} f(B).$$

Next, we must check that a_i' is generic in V over $A_i f(B)$. First, as α is generic below $\mathrm{rad}(V)$ over AB and is fixed by f, it is also generic below $\mathrm{rad}(V)$ over $A f(B)$, so there is no $A_i f(B)$-definable sub-ball of V containing u. Likewise, $u \notin \mathrm{acl}(A_i f(B))$. Thus, it suffices to check

$$a_i' \text{ is generic in } u \text{ over } A_i f(B)s. \tag{$*$}$$

So suppose not. Then there is an open sub-ball u' of u of radius $\alpha = \mathrm{rad}(u)$, algebraic over $A_i f(B)s$ and containing a_i'. Now $a_i' \in \mathrm{acl}(A_i a_i)$, so $u' = B_{<\alpha}(a_i') \in \mathrm{acl}(A_i a_i s)$. Since $a_i \underset{\mathrm{acl}(A_i s)}{\overset{g}{\downarrow}} f(B)$, we have $a_i \underset{\mathrm{acl}(A_i s)}{\overset{g}{\downarrow}} u'$. The easy Lemma 2.5.3 of [12] now yields that $u' \in \mathrm{acl}(A_i s)$. Thus, if $b_1, b_2 \in K$ are in the ball s with $b_1 \in u'$ and $b_2 \notin u'$, then $b_1 \not\equiv_{A_i s} b_2$. By [12, Corollary 2.4.5(ii)], this contradicts the fact that $x \in V$ determines a complete type over A_i.

Finally, let $j > i$. We first show that a_j is generic in U_j over $A_j\alpha$. Suppose this is false. Observe that U_j is not closed, as otherwise there would be an element of the strongly minimal set $\mathrm{red}(U_j)$ which is algebraic over $A_j\alpha$ but not over A_j, which is impossible by the instability of Γ. There is an $A_j\alpha$-definable proper sub-ball of U_j, containing a_j; hence, by Proposition 7.23, there is a proper sub-ball s_j of U_j definable over A_j. It follows by Lemma 7.29 that $\Gamma(A_j a_j) \neq \Gamma(A_j)$, contradicting the fact that $\Gamma(C) = \Gamma(A)$. Hence $a_j \underset{A_j}{\overset{g}{\cupdot}} \alpha$. Thus, as $s \in \mathrm{dcl}(A_j\alpha)$, $a_j \underset{A_j}{\overset{g}{\cupdot}} \hat{C}$ so by the assumption on f, $a_j \underset{A_j}{\overset{g}{\cupdot}} f(B)$, proving the claim.

(ii) Since $a \underset{C}{\overset{g}{\cupdot}} f(B)$, the only point to check is that $a'_i \underset{\hat{C}A_i}{\overset{g}{\cupdot}} f(B)$. This was proved in (*) above.

(iii) Since $a \underset{C}{\overset{g}{\cupdot}} B$, the only point to check is that $a'_i \underset{A_i}{\overset{g}{\cupdot}} B$. Since $a_i \underset{A_i}{\overset{g}{\cupdot}} B$, this follows from Lemma 14.9.

By parts (i) and (iii) of the claim, $AB \equiv_C Af(B)$. Thus, if $A \underset{C}{\overset{m}{\cupdot}} f(B)$ then also $A \underset{C}{\overset{m}{\cupdot}} B$. Thus, we may replace $f(B)$ by B, that is, we may assume that $A \underset{C}{\overset{g}{\cupdot}} B$ via a' and $A \underset{\hat{C}}{\overset{g}{\cupdot}} B$ via a'.

Choose a_{n+1}, b_1, b_2 generically in u to ensure that if $\hat{A} := \mathrm{acl}_K(Aa_{n+1})$, $\hat{B} := \mathrm{acl}_K(Bb_1b_2)$, then $\hat{A} \underset{C}{\overset{g}{\cupdot}} \hat{B}$ via the sequence $(a_1, \ldots, a_{i-1}, a'_i, a_{i+1}, \ldots, a_n, a_{n+1})$. Then $\hat{A}$ and $\hat{B}$ are resolved. As u is definable over $a'_i a_{n+1}$ and over $b_1 b_2$, we have $C_2 \subset \mathrm{dcl}(A_2) \cap \mathrm{dcl}(B_2)$. Also, $\Gamma(C_2) = \Gamma(A_2)$: for $\Gamma(C) = \Gamma(A)$, $\Gamma(C) \neq \Gamma(C_2)$, and $\mathrm{trdeg}(A_2/A) \leq 1$ so $\mathrm{rk}_{\mathbb{Q}}(\Gamma(A_2)/\Gamma(A)) = 1$. Thus, the hypotheses of the theorem are satisfied with C_2 in place of C, $(a_1, \ldots, a_{i-1}, a'_i, a_{i+1}, \ldots, a_n, a_{n+1})$ in place of a, and A_2, B_2 in place of A, B. Furthermore, δ has been reduced: for a'_i, a_{n+1} were chosen generically in a closed ball (namely u), and for each $j \notin \{i, n+1\}$, $a_j \underset{C_2 A_i}{\overset{g}{\cupdot}} B$. It follows that $\mathrm{tp}(A_2/C_2) \cup \mathrm{tp}(B_2/C_2) \vdash \mathrm{tp}^+(A_2 B_2/C_2)$, so $\mathrm{tp}(A/C_2) \cup \mathrm{tp}(B/C_2) \vdash \mathrm{tp}^+(AB/C_2)$.

It remains to replace C_2 by C in this conclusion, so we must show $Ah(B)$ satisfies $\mathrm{tp}^+(AB/C)$. Thus, we must show $B \equiv_{C_2} h(B)$. Recall that α was chosen generically below $\mathrm{rad}(V)$ over $CABh(B))$. In particular, $B \equiv_{C\alpha} h(B)$.

Suppose first that B (and hence also $h(B)$) has no element in V. Then $x \in V$ determines a complete type over B, and so $a'_i \underset{C}{\overset{g}{\cupdot}} B$, $a'_i \underset{C}{\overset{g}{\cupdot}} h(B)$. By Corollary 8.13, $B \equiv_{\mathrm{acl}_K(Ca'_i)} h(B)$. As α is generic below $CABh(B)$, it is generic below $\mathrm{rad}(V)$ over $\mathrm{acl}_K(Ca'_i)B$ and over $\mathrm{acl}_K(Ca'_i)h(B)$, so $B \equiv_{\alpha\,\mathrm{acl}_K(Ca'_i)} h(B)$. It follows by Lemma 7.26 that $B \equiv_{C_2} h(B)$ in this case.

Finally, suppose that there is $b \in B \cap V$. Then, as $a'_i, b, h(b) \in V$, $\alpha > |a'_i - b|$ (so $b \in u$), and also $\alpha > |a'_i - h(b)|$. Choose a field element d generically in the ball u over $CBh(B)$. Then $|d - b| = \alpha$ and $|d - h(b)| = \alpha$. Also, by the choice of α, $d \underset{C}{\overset{g}{\cupdot}} B$ and $d \underset{C}{\overset{g}{\cupdot}} h(B)$. Put $D := \mathrm{acl}_K(Cd)$. Then by Corollary 8.13, $B \equiv_D h(B)$. Since $\alpha = |d - b| = |d - h(b)|$, $B \equiv_{D\alpha} h(B)$. Hence, by Lemma 7.26, $B \equiv_{\mathrm{acl}(D\alpha)} h(B)$. Since $u = B_{\leq\alpha}(d)$, $B \equiv_{C_2} h(B)$, as required. □

Using results from this and the last chapter, we summarise how $\downarrow^g$ and $\downarrow^m$ behave, given some orthogonality to Γ. First we need a lemma.

LEMMA 14.10. *Let* $a = (a_1, \ldots, a_n) \in K^n$, *and let* $C \leq A$ *be algebraically closed valued fields, with* $A = \mathrm{acl}_K(Ca)$ *and* $\Gamma(C) = \Gamma(A)$. *Within the ring* $C[a]$ *there are* $e_1, \ldots, e_m \in R$ *algebraically independent over* C *such that* $\{\mathrm{res}(e_1), \ldots, \mathrm{res}(e_m)\}$ *is a transcendence basis of* $k(A)$ *over* $k(C)$.

PROOF. Choose $d_1, \ldots, d_m \in A$ such that $\{\mathrm{res}(d_1), \ldots, \mathrm{res}(d_m)\}$ is a transcendence basis of $k(A)$ over $k(C)$. For each $i = 1, \ldots, m$, as d_i is algebraic over $C(a)$, there is a polynomial f_i over $C[a]$ with $f_i(d_i) = 0$. As $\Gamma(C) = \Gamma(Ca)$, we may arrange (by multiplying f_i by a scalar from C) that the maximum of the absolute values of the coefficients of f_i is 1. Thus, the reduction of f_i modulo $\mathcal{M}$ is a non-zero polynomial over $k(C(a))$, so $\mathrm{res}(d_i)$ is algebraic over the residues of the coefficients of f_i. Let E be the set of all the coefficients of all the f_i. Then each $\mathrm{res}(d_i)$ is algebraic over $\{\mathrm{res}(x) : x \in E\}$. Hence, there are $e_1, \ldots, e_m \in E$ such that $\{\mathrm{res}(e_1), \ldots, \mathrm{res}(e_m)\}$ is a transcendence basis of $k(A)$ over $k(C)$. Then $e_1, \ldots, e_m$ are algebraically independent over C. □

THEOREM 14.11. *Let* $C \leq A, B$ *be algebraically closed valued fields with* A, B *extensions of* C *of finite transcendence degree, and at least one of* $\mathrm{tp}(A/C)$, $\mathrm{tp}(B/C)$ *orthogonal to* Γ. *Then the following are equivalent.*

  (i) $A \downarrow^m_C B$.

  (ii) $A \downarrow^g_C B$ *via some generating sequence.*

  (iii) $A \downarrow^g_C B$.

  (iv) $B \downarrow^g_C A$ *via some generating sequence.*

  (v) $B \downarrow^g_C A$.

  (vi) $A \downarrow^d_C B$.

PROOF. We assume $\mathrm{tp}(A/C) \perp \Gamma$. Then (ii) ⇒ (i) is Theorem 14.4, and we prove (i) ⇒ (ii). If instead $\mathrm{tp}(B/C) \perp \Gamma$, then the same argument proves (i) ⇔ (iv). By Proposition 10.11, we have the equivalence of (ii), (iii), (iv), (v), (vi).

So assume $A \downarrow^m_C B$. By Lemma 14.10, there are $e_1, \ldots, e_m \in A$ algebraically independent over C such that $\mathrm{res}(e_1), \ldots, \mathrm{res}(e_m)$ form a transcendence basis of $k(A)$ over $k(C)$. As $A \downarrow^m_C B$, also $C[e_1, \ldots, e_m] \downarrow^m_C B$, and it follows that $\mathrm{res}(e_1), \ldots, \mathrm{res}(e_m)$ are linearly independent (so algebraically independent) over $k(B)$. For suppose that $\Sigma_{i=1}^m \mathrm{res}(b_i)\,\mathrm{res}(e_i) = 0$. Then $|\Sigma_{i=1}^m b_i e_i| < 1$, so $\mathrm{tp}(b_1, \ldots, b_m/C) \cup \mathrm{tp}(e_1, \ldots, e_m/C) \vdash |\Sigma_{i=1}^m x_i y_i| < 1$. Hence $\mathrm{res}(b_1) = \cdots = \mathrm{res}(b_m) = 0$: indeed, suppose say $\mathrm{res}(b_1) \neq 0$. Let

e_1' be chosen generically in R over all other parameters. Then $e_1'e_2 \ldots e_m \equiv_C$ $e_1 \ldots e_m$, so $|e_1'b_1 + \Sigma_{i=2}^m e_i b_i| < 1$, but

$$|e_1'b_1 + \Sigma_{i=2}^m e_i b_i| = |(e_1' - e_1)b_1 + \Sigma_{i=1}^m e_i b_i| = |(e_1' - e_1)b_1| = 1,$$

a contradiction.

Hence, $k(A) \underset{C}{\overset{g}{\bigcup}} B$. As C is resolved, it follows that $\mathrm{St}_C(A) \underset{C}{\bigcup} \mathrm{St}_C(B)$. Hence, by Proposition 10.11, $A \underset{C}{\overset{g}{\bigcup}} B$ via some (in fact, every) generating sequence. □

THEOREM 14.12. *Let* $C \subseteq \mathcal{U}$, *and let* V *be an affine variety defined over* $C \cap K$. *Let* $p|C$ *be a stably dominated type over* C *of elements of* V, *with* $\mathrm{Aut}(\mathcal{U}/C)$-*invariant extension* p. *Let* F *be a regular function on* V, *defined over a field* L *with* $C \subseteq \mathrm{dcl}(L)$. *Then* $|F(x)|$ *has a maximum* $\gamma_{\max}^F \in \Gamma(C)$ *on* $\{x \in \mathcal{U} : x \models p|C\}$.

Moreover for $a \models p|C$, *we have:* $a \models p|L$ *if and only if* $|F(a)| = \gamma_{\max}^F$ *for all such* F.

PROOF. Since $p|C$ has an $\mathrm{Aut}(\mathcal{U}/C)$-invariant extension, it implies $p|\,\mathrm{acl}(C)$; so with no loss of generality we may assume $C = \mathrm{acl}(C)$.

Choose an embedding of V in $\mathbb{A}^n$. Then F lifts to a polynomial on $\mathbb{A}^n$. Thus we may assume $V = \mathbb{A}^n$. Write $F = F(x, b)$. If $a \models p|L$ then $\Gamma(La) = \Gamma(L)$. So for some $\alpha \in \Gamma(L)$, if $a \models p|L$ then $|F(a, b)| = \alpha$. The hypotheses of Corollary 14.6 hold. Hence $p(x)|C \vdash |F(x, b)| \leq \alpha$. This shows that α is the maximum of $|F|$ on the realizations of $p|C$.

The 'moreover' follows from quantifier elimination, since the set of formulas $\{|F(x)| = \gamma_{\max}^F : F(X) \in L[X]\}$ determines a complete type over L. □

Here is a two-sided version of the same result. If $p(x), q(y)$ are types, define $p \times q(x, y) := p(x) \cup q(y)$. If p is stably dominated, let $p \otimes q$ be the complete type with realisations

$$\{(a, b) : (a, b) \models p \otimes q \text{ and } a \underset{C}{\overset{d}{\bigcup}} b\}.$$

THEOREM 14.13. *Let* $F(x, y)$ *be a polynomial over the algebraically closed valued field* C, *and* p, q *be stably dominated types in the field sort over* C. *Assume* p *is stably dominated. Then* $|F(x, y)|$ *has a maximum* $\gamma_{\max} \in \Gamma(C)$ *on* $p \times q$. *Also,*

$$(a, b) \models p \otimes q \Rightarrow |F(a, b)| = \gamma_{\max}.$$

PROOF. First, if $(a, b) \models p \otimes q$ then $\Gamma(Cab) = \Gamma(b) = \Gamma(C)$, by stable domination (twice). Thus, $|F(x, y)|$ takes constant value in $\Gamma(C)$. Now apply Theorem 14.4. □

Here is a group-theoretic consequence of the maximum modulus principle. If G is an affine algebraic group over K, we have the ring of regular functions $K[G]$. We can also view G as a definable group in ACVF. Recall the discussion of translation invariant definable types above 6.13.

THEOREM 14.14. *Let G be an affine algebraic group over a field C, and H a C-definable subgroup of $G(K)$. Let p be a C-definable global type of elements of H. Then the following are equivalent:*

(i) *p is translation invariant, and stably dominated.*

(ii) *For any $f \in C[G]$, $b \models p$, and $a \in H$ then $|f(a)| \le |f(b)|$.*

PROOF. (i) $\Rightarrow$ (ii) Since $p|C \perp \Gamma$, $|f(x)|$ takes a fixed value γ_f on $p|C$. If $(a, b) \models p \otimes p$ then $ab \models p$ (cf. Section 3 of [15]). In particular, $|f(ab)| = \gamma_f$. Also, it follows from Theorem 14.13 that for $(a, b) \models p \times p$, $|f(ab)| \le \gamma_f$. Now $H = \{ab : (a, b) \models p \times p\}$.

(ii) $\Rightarrow$ (i) By quantifier elimination in ACVF in the language L_{div} (see Theorem 7.1), condition (2) characterises p uniquely. Also, condition (2) is translation-invariant, so the type characterised is translation-invariant. Clearly $\gamma_f \in \Gamma(C)$ for each f. It follows by quantifier elimination and Proposition 12.5 that if $C' \supseteq C$ is a maximally complete algebraically closed valued field then $p|C'$ is stably dominated, so $p|C$ is stably dominated by Theorem 4.9.   □

The equivalence implies in particular the uniqueness of stably dominated, translation-invariant types; this can also be seen abstractly. The conditions (i), (ii) hold for example if $G = \mathrm{SL}_n(K)$, $H = \mathrm{SL}_n(R)$, and p is the unique type of H whose elements have reduction satisfying the generic type of $\mathrm{SL}_n(k)$. See [15] for more detail.

CHAPTER 15

CANONICAL BASES AND INDEPENDENCE GIVEN BY MODULES

Consider the following situation in a pure algebraically closed field K. Suppose $C \leq A \cap B$ with C algebraically closed. Let p be a type, and put

$$I(p) = \{f \in K[X]; \text{ for all } a \models p \ (f(a) = 0)\}.$$

Then $A \underset{C}{\downarrow} B$ in the sense of stability theory if and only if $I(a/C) = I(a/B)$ for all tuples a from A. We refine this to take the valuation into account, and use it to define another notion of independence. We will show that some orthogonality to Γ is enough to prove its equivalence with the other forms of independence.

In this chapter, for ease of notation, if $s \in S_n \cup T_n$ we identify s with the subset of K^n which it codes.

DEFINITION 15.1. Let p be a partial field type over some set of parameters. Define

$$J(p) := \{f(X) \in K[X]: p(x) \vdash |f(x)| < 1\}.$$

More generally, if $p = p(x, v)$ is a partial type with $x = (x_1, \ldots, x_\ell)$ field variables and $v = (v_1, \ldots, v_m)$ a tuple of variables with v_i ranging through $S_{n_i} \cup T_{n_i}$, then

$$J(p) = \{f(X, Y) \in K[X, Y]:$$
$$p(x, v) \vdash (\forall y_1 \in v_1) \ldots (\forall y_m \in v_m)|f(x, y)| < 1\}.$$

Here, y_i is an n_i-tuple of field variables for each $i = 1, \ldots m$. The notation $y_i \in v_i$ is a slight abuse – it means that y_i lies in a lattice (or in an element of some $\mathrm{red}(s)$) *coded* by an element of $S_{n_i} \cup T_{n_i}$. We shall often write $J(a/B)$ for $J(\mathrm{tp}(a/B))$.

Next, for a tuple a in $\mathcal{U}$, and B, C lying in arbitrary sorts, define $a \underset{C}{\overset{J}{\downarrow}} B$ to hold if and only if $J(a/C) = J(a/BC)$. Finally, if $A \subseteq \mathcal{U}$, then $A \underset{C}{\overset{J}{\downarrow}} B$ holds if $a \underset{C}{\overset{J}{\downarrow}} B$ for all finite tuples a from A.

161

If $J_n(p) := \{f \in J(p): \deg(f) < n\}$ (where deg refers to total degree), then $J_n(p)$ can be regarded as an R-submodule of K^N for some N: identify each polynomial with the correspondinq sequence of coefficients. Note that $J_n(p)$ is the union of a collection of *definable* R-modules (one module for each formula in p). For if $p = \{\varphi_i: i \in I\}$, let

$$J_n(\varphi) := \{f(X) \in K[X]: \forall x(\varphi(x) \rightarrow |f(x)| < 1\},$$

a definable R-module. Then an easy compactness argument shows that $J_n(p) = \bigcup(J_n(\varphi_i): i \in I)$.

In the above, we think of K as the field sort of $\mathcal{U}$, so $K[X]$ denotes $(K \cap \mathcal{U})[X]$. Observe that we do not need to consider in addition the polynomials satisfying a weak inequality. For suppose $J(a/C) = J(a'/C)$, and $\text{tp}(a/C) \vdash |f(x)| \leq 1$. Then for any $\delta \in \Gamma(K)$ with $\delta > 1$ there is $d \in K$ with $|d| = \delta$. Now $d^{-1}f(x) \in J(a/C) = J(a'/C)$, so for all $a'' \equiv_C a'$, $|f(a'')| < \delta$. It follows by saturation of K that for all such a'', $|f(a'')| \leq 1$, that is, $\text{tp}(a'/C) \vdash |f(x)| \leq 1$.

For an invariant type p, $J(p)$ is defined as for types over a fixed set; here however we can write more simply:

$$J(p) = \{f(X) \in K[X]: (|f(x)| < 1) \in p\}.$$

Note that if p is a definable type, then $J(p)$ is *definable*, in the sense that the intersection with the space $K[X]_d$ of polynomials of degree $\leq d$ is definable, for any integer d.

If p is a stably dominated type of field elements over $C = \text{acl}(C) \subseteq \mathcal{U}$, and p' is the corresponding invariant type, the maximum modulus principle (in the form of Theorem 14.12) states precisely that $J(p) = J(p')$.

When p is stably dominated, we will see that knowldege of $J(p)$ determines p. Thus the sequence of codes for the submodules $(K[x]_d \cap J(p))$ can be viewed as the *canonical base* of p. This can be seen as giving geometric meaning to the canonical base of a stably dominated type, analogous to the field of definition of a prime ideal in ACF.

For J-independence we have transitivity on the right, and unlike with $\underset{}{\overset{g}{\downarrow}}$, there is no dependence on a generating sequence. On the other hand transitivity on the left fails (as follows from Example 15.12, along with Proposition 15.10); hence symmetry can fail. Existence of J-independent extensions of $\text{tp}(A/C)$ can also fail (Example 15.13), though, by Theorem 15.5 below, it holds if $\text{tp}(A/C) \perp \Gamma$. We show below that, given some orthogonality to Γ, $\underset{}{\overset{J}{\downarrow}}$ coincides with $\underset{}{\overset{g}{\downarrow}}$, and so has these properties. There are further results and examples at the end of the chapter.

LEMMA 15.2. *Suppose $C \leq B$ are valued fields with C algebraically closed, and $a \in K^n$ with $a \underset{C}{\overset{J}{\downarrow}} B$. Then $a \underset{C}{\overset{m}{\downarrow}} B$.*

PROOF. Suppose that $\mathrm{tp}(ab/C) \vdash |f(x_a, y_b)| < \gamma$ where b is a tuple in B, and f is over $\mathbb{Z}$. We may suppose that $\gamma \in \Gamma(C)$. Choose $d \in K$ with $|d| = \gamma$. We need to show that $\mathrm{tp}(a/C) \vdash |f(x, b)| < \gamma$. Now $\mathrm{tp}(a/B) \vdash |f(x, b)| < \gamma$, so $d^{-1}f(x, b) \in J(a/B) = J(a/C)$. Hence, if $a' \equiv_C a$, then $|f(a', b)| < \gamma$.

Next, suppose $\mathrm{tp}(ab/C) \vdash |f(x, y)| \leq \gamma$. Let $\delta \in \Gamma$ with $\gamma < \delta$. Then $\mathrm{tp}(ab/C) \vdash |f(x, y)| < \delta$. So, arguing as above, if $a' \equiv_C a$ then $|f(a', b)| < \delta$. Thus, if $a' \equiv_C a$ then $|f(a', b)| \leq \gamma$. $\qquad\square$

Notice that the converse of the above lemma is not obviously true. This is because, in the definition of J-independence, the coefficients of the polynomials are allowed to be anywhere in K, whereas modulus independence only refers to polynomials with parameters from the given set. In the proof of the following lemma, we see a way of overcoming this issue. Then we show that sequential independence implies J-independence, given some orthogonality to Γ.

LEMMA 15.3. *Suppose* $C = \mathrm{acl}(C) \subseteq A \cap B$ *with* $\Gamma(C) = \Gamma(A)$, A *is resolved, and* $A \cap K \underset{C}{\overset{J}{\smile}} B$. *Then* $A \underset{C}{\overset{J}{\smile}} B$.

PROOF. For ease of notation, we just do the following special case: $t = (t_1 \ldots, t_m)$ is a tuple of elements of $\mathrm{red}(s)$, where $s \in S_n$, and $f(U) \in J(t/B)$. We must show that $f(U) \in J(t/C)$. As A is resolved, there is $a_i \in K^n \cap t_i \cap A$ for each $i = 1, \ldots m$. Put $a = (a_1, \ldots, a_m)$. Furthermore, there is in A a basis $b = (b_1, \ldots, b_n)$ for s in K (so $b_i \in K^n \cap A^n$ for each i). Note that $t_i := \{a_i + \Sigma_{j=1}^n y_{ij}b_j : y_{ij} \in R\}$ for each i. Let $q := \mathrm{tp}(ab/B)$. Then, as $f(U) \in J(t/B)$, $q(x_a, z_b)$ implies

$$(\forall y_{ij} \in \mathcal{M}: 1 \leq i \leq m, 1 \leq j \leq n)$$
$$(|f(x_1 + \Sigma_{j=1}^n y_{1j}z_j, \ldots, x_m + \Sigma_{j=1}^n y_{mj}z_j)| < 1).$$

In particular, for any $y_{ij} \in \mathcal{M}_K$,

$$q(x_a, z_b) \vdash |f(x_1 + \Sigma_{j=1}^n y_{1j}z_j, \ldots, x_m + \Sigma_{j=1}^n y_{mj}z_j)| < 1.$$

Thus, $f(x_1 + \Sigma_{j=1}^n y_{1j}z_j, \ldots, x_m \Sigma_{j=1}^n y_{mj}z_j) \in J(q)$. By assumption, $J(q) = J(q|c)$, so $f(x_1 + \Sigma_{j=1}^n y_{1j}z_j, \ldots, x_m + \Sigma_{j=1}^n y_{mj}z_j) \in J(q|c)$. By compactness, this forces that $q|c(x_a, z_b)$ implies the first displayed inequality above, hence $f \in J(t/C)$. $\qquad\square$

It is in the proof of the next theorem that Chapter 14 is used: $\underset{}{\overset{m}{\smile}}$ provides the link from $\underset{}{\overset{g}{\smile}}$ to $\underset{}{\overset{J}{\smile}}$.

THEOREM 15.4. *Assume* $C = \mathrm{acl}((C \cap K) \cup H)$, *where* H *is a subset of* $S = \bigcup_{n>0} S_n$. *Suppose also that* $C \subseteq A \cap B$ *and* $\Gamma(C) = \Gamma(A)$. *Assume* $A \underset{C}{\overset{g}{\smile}} B$ *via a sequence of field elements. Then* $A \underset{C}{\overset{J}{\smile}} B$.

PROOF. Since, by Corollary 10.15, we may resolve A generically over B without losing the assumption $\Gamma(C) = \Gamma(A)$, and may also resolve B generically, we may assume $A = \mathrm{acl}(A \cap K)$ and $B = \mathrm{acl}(B \cap K)$. Let $f(x, e) \in J(A/B)$. Since A is resolved, we may by Lemma 15.3 assume x ranges through the field sorts. Notice that the tuple of coefficients e may be anywhere in K. We must show $f(x, e) \in J(A/C)$. Since this is a property of $\mathrm{tp}(A/C)$, and our assumption on $f(x, e)$ depends on $\mathrm{tp}(A/B)$, by translating A over B we may assume $A \underset{B}{\overset{g}{\downarrow}} e$. Hence $A \underset{C}{\overset{g}{\downarrow}} Be$. It follows by Theorem 14.4 that $\mathrm{tp}(A/C) \cup \mathrm{tp}(Be/C) \vdash \mathrm{tp}^+(A_K B_K e/C)$, and so $\mathrm{tp}(A/C) \vdash |f(x, e)| < 1$. □

Theorem 15.5 and Corollary 15.7 to follow can also be proved very rapidly using the observation $J(p) = J(p|C)$ for stably dominated p based on C, made at the beginning of the chapter. However Lemma 15.6 covers cases that cannot be seen in this way.

THEOREM 15.5. *Assume* $C = \mathrm{acl}(C)$, A, B *are $\mathcal{L}_{\mathcal{G}}$-structures, finitely acl-generated over C, and that $C \subseteq A \cap B$, with $\mathrm{tp}(A/C) \perp \Gamma$ and $A \underset{C}{\overset{g}{\downarrow}} B$. Then $A \underset{C}{\overset{J}{\downarrow}} B$.*

PROOF. We shall generically resolve C and then use the last theorem. Let C' be a closed resolution of C, and M a canonical open resolution of C'. We may choose C' and M so that $C' \underset{C}{\overset{g}{\downarrow}} AB$ and $M \underset{C'}{\overset{g}{\downarrow}} C'AB$. Note that here we make essential use of *transfinite* sequential independence. In particular, as $\mathrm{tp}(A/C) \perp \Gamma$, $A \underset{C}{\overset{g}{\downarrow}} C'$ by Proposition 10.11, so $\mathrm{tp}(A/C') \perp \Gamma$ by Lemma 10.2 (iii).

Now $\mathrm{tp}(C'/C) \perp \Gamma$, so by Proposition 10.11 $AB \underset{C}{\overset{g}{\downarrow}} C'$ via any generating sequence, and hence $A \underset{BC}{\overset{g}{\downarrow}} BC'$. As $A \underset{C}{\overset{g}{\downarrow}} B$, we obtain $A \underset{C}{\overset{g}{\downarrow}} BC'$, so $A \underset{C'}{\overset{g}{\downarrow}} BC'$. It follows that $\mathrm{tp}(A/BC') \perp \Gamma$, so as $M \underset{C'B}{\overset{g}{\downarrow}} A$ (since $M \underset{C'}{\overset{g}{\downarrow}} AB$) we obtain $A \underset{C'B}{\overset{g}{\downarrow}} M$. Since also $A \underset{C'}{\overset{g}{\downarrow}} BC'$, we have $A \underset{C'}{\overset{g}{\downarrow}} BM$. Hence, $A \underset{C}{\overset{g}{\downarrow}} BM$ (as $A \underset{C}{\overset{g}{\downarrow}} C'$), so, finally, $A \underset{M}{\overset{g}{\downarrow}} B$. In addition, as $M \underset{C'}{\overset{g}{\downarrow}} A$, and $\mathrm{tp}(A/C') \perp \Gamma$, we have $A \underset{C'}{\overset{g}{\downarrow}} M$. Thus $A \underset{C}{\overset{g}{\downarrow}} M$, so $\mathrm{tp}(A/M) \perp \Gamma$.

Put $A'' := \mathrm{acl}(AM)$ and $B'' := \mathrm{acl}(BM)$. Then $A'' \underset{M}{\overset{g}{\downarrow}} B''$ and $\mathrm{tp}(A''/M) \perp \Gamma$. Furthermore, there is a resolution A''' of A'' such that $A''' \underset{M}{\overset{g}{\downarrow}} B''$ and $\mathrm{tp}(A'''/M) \perp \Gamma$ (by Lemma 10.14). By Lemma 10.9, $A''' \underset{M}{\overset{g}{\downarrow}} B''$ via a sequence of field elements. It follows from Theorem 15.4 that $A''' \underset{M}{\overset{J}{\downarrow}} B''$, so $A'' \underset{M}{\overset{J}{\downarrow}} B''$.

To prove the theorem, suppose $f(x, e) \in J(A/B)$; that is, for some subsequence a of A, $\mathrm{tp}(A/B) \vdash |f(a, e)| < 1$ (for ease of notation – not by Lemma 15.3 – we are assuming that a is a tuple of field elements). We must show $\mathrm{tp}(A/C) \vdash |f(x, e)| < 1$, or equivalently, $\mathrm{tp}(e/C) \vdash |f(a, y)| < 1$. So suppose $e' \equiv_C e$. Since we may translate ee' over AB, we may suppose $M \underset{AB}{\overset{g}{\downarrow}} ABee'$. Also, $M \underset{C}{\overset{g}{\downarrow}} AB$, so $M \underset{C}{\overset{g}{\downarrow}} ABee'$. Thus $e \equiv_M e'$ by Corollary 8.14.

We have $f(x, e) \in J(A''/B'')$, and $A'' \underset{M}{\overset{J}{\downarrow}} B''$, so $\mathrm{tp}(A''/M) \vdash |f(x, e)| < 1$. Since x refers just to elements from A, $\mathrm{tp}(A/M) \vdash |f(x, e)| < 1$, so $\mathrm{tp}(e/M) \vdash |f(a, y)| \leq 1$. Hence, $|f(a, e')| < 1$. $\qquad \square$

By the use of the following lemma, we can prove the analogue of Theorem 15.5, but with the orthogonality condition on the right.

LEMMA 15.6. *Suppose* $C = \mathrm{acl}(C \cap K)$ *and* $C \subseteq A \cap B$. *Suppose too that* A *is resolved,* $A \underset{C \cup \Gamma(A)}{\overset{g}{\downarrow}} B$ *via a sequence of field elements, and that* $\mathrm{tp}(\Gamma(A)/C) \vdash \mathrm{tp}(\Gamma(A)/C\Gamma(B))$. *Then* $A \underset{C}{\overset{J}{\downarrow}} B$.

PROOF. Let $A_1 := C \cup \Gamma(A)$. Since $\Gamma(A)$ consists of elements of S_1, $A \underset{A_1}{\overset{J}{\downarrow}} B \cup \Gamma(A)$ by Theorem 15.4. Suppose now $f(X, Y) \in \mathbb{Z}[X, Y]$, and $e \in K^n$ with $f(x, e) \in J(A/B)$, so $\mathrm{tp}(A/B) \vdash |f(x, e)| < 1$ (so x corresponds to some subtuple of A). We must show $\mathrm{tp}(A/C) \vdash |f(x, e)| < 1$. As usual, we may take x in the field sorts by Lemma 15.3.

CLAIM. $\mathrm{tp}(A/C) \cup \mathrm{tp}(A_1/B) \cup \mathrm{tp}(e/B) \vdash |f(x, y)| < 1$.

PROOF. Fix A_1 and B, and suppose $e' \equiv_B e$. We must show $\mathrm{tp}(A/A_1) \vdash |f(x, e')| < 1$. Suppose now that $A' \equiv_{A_1} A$, with $A' \underset{A_1}{\overset{g}{\downarrow}} Be'$ via the corresponding sequence of field elements. Then $\mathrm{tp}(A'B) = \mathrm{tp}(AB)$ (since also $A \underset{A_1}{\overset{}{\downarrow}} B$), so $\mathrm{tp}(A'/B) \vdash |f(x, e')| < 1$ (as $\mathrm{tp}(A/B) \vdash |f(x, e)| < 1$). By Theorem 15.4 (since $\Gamma(A) \subset S_1$), $A' \underset{A_1}{\overset{J}{\downarrow}} Be'$, so $\mathrm{tp}(A'/A_1) \vdash |f(x, e')| < 1$. As $A' \equiv_{A_1} A$, this proves the claim.

Now by assumption $\mathrm{tp}(A_1/C) \vdash \mathrm{tp}(A_1/C\Gamma(B))$, so $\mathrm{tp}(A_1/C) \vdash \mathrm{tp}(A_1/B)$ (as Γ is stably embedded and $A_1 \subset C \cup \Gamma$). Hence, $\mathrm{tp}(A/C) \vdash |f(x, e)| < 1$, as required. $\qquad \square$

COROLLARY 15.7. *Suppose* $C = \mathrm{acl}(C) \subseteq A \cap B$ *with* A, B *finitely acl-generated over* C, *and that* $\mathrm{tp}(B/C) \perp \Gamma$ *and* $A \underset{C}{\overset{g}{\downarrow}} B$. *Then* $A \underset{C}{\overset{J}{\downarrow}} B$.

PROOF. Since we may replace A by a generic resolution over B, we may suppose that A is resolved. By Proposition 10.11, $B \underset{C}{\overset{g}{\downarrow}} A$. As in the proof of Theorem 15.5, let C' be a generic closed resolution of C, and M be a canonical open resolution of C', with $C' \underset{C}{\overset{g}{\downarrow}} AB$ and $M \underset{C'}{\overset{g}{\downarrow}} C'AB$. Put $A'' := \mathrm{acl}(AM)$ and $B'' := \mathrm{acl}(BM)$, so as in 15.5 we have $B'' \underset{M}{\overset{g}{\downarrow}} A''$ and $\mathrm{tp}(B''/M) \perp \Gamma$. Then $B'' \underset{M \cup \Gamma(A'')}{\overset{g}{\downarrow}} A''$. Hence, as $\mathrm{tp}(B''/M \cup \Gamma(A'')) \perp \Gamma$, we have $A'' \underset{M \cup \Gamma(A'')}{\overset{g}{\downarrow}} B''$ via a sequence of field elements. Now apply Lemma 15.6, noting that $\Gamma(B'') = \Gamma(M)$ as $\mathrm{tp}(B/M) \perp \Gamma$, to obtain $A'' \underset{M}{\overset{J}{\downarrow}} B''$. The last part of the proof of Theorem 15.5 now shows $A \underset{C}{\overset{J}{\downarrow}} B$. $\qquad \square$

We have the following strong converse in any sorts, only assuming some orthogonality to Γ.

THEOREM 15.8. *Let* $A = \mathrm{dcl}(A)$, $B = \mathrm{acl}(B)$, $C = \mathrm{acl}(C)$. *Suppose* $A \underset{C}{\overset{J}{\bigcup}} B$, *and that* $\mathrm{tp}(A/C) \perp \Gamma$, *or* $\mathrm{tp}(B/C) \perp \Gamma$. *Then* $A \underset{C}{\overset{g}{\bigcup}} B$.

REMARK. 1. Since one of $\mathrm{tp}(A/C), \mathrm{tp}(B/C)$ is orthogonal to Γ, the notion of independence in the conclusion is symmetric and independent of a choice of generating set. Furthermore, if $\mathrm{tp}(A/C) \perp \Gamma$, then the conclusion yields $\Gamma(AB) = \Gamma(B)$, and if $\mathrm{tp}(B/C) \perp \Gamma$, we obtain $\Gamma(AB) = \Gamma(A)$.

2. In the proof below, we use Srour's notion of an *equation* over a set B, that is, a formula $\varphi(x)$ such that the intersection of any set of conjugates of $\varphi(\mathcal{U})$ over B is equal to a finite sub-intersection (see e.g., [52]). We shall use results from [41] about equations in stable theories, applied within the stable structure St_C.

PROOF. It suffices to show $\mathrm{St}_C(A) \bigcup \mathrm{St}_C(B)$. For by Proposition 10.11, under either orthogonality hypothesis this yields $A \underset{C}{\overset{g}{\bigcup}} B$.

Let $t = (t_1, \ldots, t_m) \in \mathrm{St}_C(A)$. We may suppose $m = 1$ and $t \in \mathrm{red}(s)$ where $s \in \mathrm{dcl}(C) \cap S_\ell$; for if $t_i \in \mathrm{res}(s_i)$, where $s_i \in \mathrm{dcl}(C)$, then $(t_1, \ldots, t_m)$ can be regarded as an element of $\mathrm{red}(\Lambda(s_1) \times \cdots \times \Lambda(s_m))$, and $\Lambda(s_1) \times \cdots \times \Lambda(s_m)$ is a C-definable lattice. We must show $t \underset{C}{\bigcup} \mathrm{St}_C(B)$, that is, $\mathrm{RM}(t/C) = \mathrm{RM}(t/\mathrm{St}_C(B))$ in the structure St_C. Let M be a model containing B, with $t \underset{B}{\overset{g}{\bigcup}} M$. Then by Proposition 8.19, $t \underset{B}{\bigcup} \mathrm{St}_B(M)$ in St_B, so $t \underset{\mathrm{St}_C(B)}{\bigcup} \mathrm{St}_C(M)$ in St_C (compare Proposition 3.22, (i) $\Rightarrow$ (iii)); hence $\mathrm{RM}(t/\mathrm{St}_C(B)) = \mathrm{RM}(t/\mathrm{St}_C(M))$ in St_C. Thus, our task is to show $\mathrm{RM}(t/C) = \mathrm{RM}(t/\mathrm{St}_C(M))$ in St_C. As M is a model, there is an M-definable R-module isomorphism $\psi : \Lambda(s) \to R^\ell$ (an affine map) inducing an isomorphism (also denoted ψ) $\mathrm{red}(s) \to k^\ell$. Put $t' = (t_1', \ldots, t_\ell') := \psi(t) \in k^\ell$.

Let P_C (respectively P_M) denote the solution sets of $\mathrm{tp}(t/C)$ (respectively $\mathrm{tp}(t/M)$), and let Z_C (respectively Z_M) be the Zariski closure of $\psi(P_C)$ (respectively $\psi(P_M)$) in k^ℓ. We have $P_C \supseteq P_M$ and $Z_C \supseteq Z_M$. Now $\mathrm{RM}(P_C) = \mathrm{RM}(\psi(P_C)) = \mathrm{RM}(Z_C)$, and similarly $\mathrm{RM}(P_M) = \mathrm{RM}(Z_M)$. Thus, we must show $Z_C = Z_M$, as this gives $\mathrm{RM}(P_C)) = \mathrm{RM}(P_M)$.

So suppose $f(X) \in k(M)[X]$, with $f(t') = 0$. We wish to show that f vanishes on Z_C. Lift f to a polynomial $F(X)$ over $R \cap M$. Then if $y_1 \in t_1', \ldots, y_\ell \in t_\ell'$ and $y = (y_1, \ldots y_\ell) \in K^\ell$ we have $|F(y)| < 1$. Put $H(X) := F(\psi(X_1, \ldots, X_\ell))$. Then $|H(x_1, \ldots, x_\ell)| < 1$ for any $x = (x_1, \ldots, x_\ell) \in t$.

CLAIM. If u is a single variable ranging over elements of T_ℓ, then $\mathrm{tp}(t/B) \vdash f(\psi(u)) = 0$.

PROOF OF CLAIM. The formula $f(\psi(u)) = 0$ is over M: write it as $\delta(u, m)$. Working over B, $\delta(u, m)$ is an equation in the sense of Srour [52]. For if we fix an arbitrary affine bijection $\mathrm{red}(s) \to k^\ell$ (not over B), then conjugates over B of the formula $f(\psi(u)) = 0$ become formulas $g(x) = 0$ over k, and the ideal in $k[X]$ generated by any set of these is finitely generated. Furthermore, the formula $f(\psi(u)) = 0$ can be interpreted in the stable structure St_B, since this structure is stably embedded. That is, we may suppose m is from M^{st}. Also, $t \underset{B}{\perp} M$ and $f(\psi(t)) = 0$. Let X be the intersection of the solution sets of all the formulas $\delta(u, m')$ such that $m' \in M^{\mathrm{st}}$ and $\models \delta(t, m')$ holds. Then by Proposition 4.2 of [41], using that $B = \mathrm{acl}(B)$, we find that X is B-definable. The claim follows.

It follows that

$$\mathrm{tp}(t/B) \vdash \forall x \in u(|H(x_1, \ldots, x_\ell)| < 1).$$

Thus, $H \in J(t/B)$. But $J(t/B) = J(t/C)$, so

$$\mathrm{tp}(t/C) \vdash \forall x \in u(|H(x_1, \ldots, x_\ell)| < 1).$$

Hence, $\mathrm{tp}(t/C) \vdash f(\psi(u)) = 0$. It follows that f vanishes on Z_C, as required.

$\square$

We summarise the equivalences of different notions of independence, under an assumption of orthogonality to Γ.

THEOREM 15.9. *Let A, B, C be algebraically closed $\mathcal{L}_G$-structures, with $C \leq A \cap B$ and A, B finitely acl-generated over C, and suppose that $\mathrm{tp}(A/C)$ is orthogonal to Γ. Then the following are equivalent.*

(i) $A \underset{C}{\overset{J}{\perp}} B$

(ii) $A \underset{C}{\overset{g}{\perp}} B$

(iii) $B \underset{C}{\overset{g}{\perp}} A$

(iv) $B \underset{C}{\overset{J}{\perp}} A$

(v) $A \underset{C}{\overset{d}{\perp}} B$.

If all structures are valued fields, then all of the above are equivalent to

(vi) $A \underset{C}{\overset{m}{\perp}} B$.

PROOF. We have (ii) $\Leftrightarrow$ (iii) and (ii) $\Leftrightarrow$ (v) by Proposition 10.11, and parts (i) $\Rightarrow$ (ii) and (iv) $\Rightarrow$ (iii) are from Theorem 15.8. For (ii) $\Rightarrow$ (i) see Theorem 15.5, and (iii) $\Rightarrow$ (iv) comes from Corollary 15.7. Finally, for the equivalence of (vi) in the case of fields, see Theorem 14.11. $\square$

We conclude with various examples. These illustrate that sequential independence and J-independence are not equivalent in general. In all such examples, stable domination of course fails. It is worth noting that for singletons in the field sort, the notions are in fact equivalent, as we first prove below. Thus the examples must all involve at least two field elements, and there cannot be any orthogonality to the value group.

PROPOSITION 15.10. *Let $C = \mathrm{acl}(C) \subseteq B$ and $a \in K$ be a single element. Then $a \underset{C}{\overset{J}{\perp}} B$ if and only if $a \underset{C}{\overset{g}{\perp}} B$.*

PROOF. Suppose first that $a \not\perp_C^g B$. We suppose that $\mathrm{tp}(a/C)$ is the generic type of a unary set U (a ball, or the intersection of a sequence of balls, or K itself). Then there is a B-definable proper sub-ball V of U containing a. If U is a closed ball, and V is an open ball of the same radius γ, choose $d \in K$ with $|d| = \gamma$. Then $d^{-1}(x - a) \in J(a/B) \setminus J(a/C)$, so $a \not\perp_C^J B$; indeed, if $a' \equiv_B a$ then $a' \in V$, so $|x - a| < |d|$, but there is $a' \equiv_C a$ with $a' \notin V$, and then $|x - a'| = |d|$. In all the other cases for U, $\mathrm{rad}(V) < \mathrm{rad}(U)$. Now choose $d \in K$ with $\mathrm{rad}(V) < |d| < \mathrm{rad}(U)$. Then again, $d^{-1}(x - a) \in J(a/B) \setminus J(a/C)$.

For the other direction, assume $a \perp_C^g B$. Let U be a C-unary set such that $\mathrm{tp}(a/C)$ is the generic type of U. Let $f(x) \in J(a/B)$. We must show that $f \in J(a/C)$. So suppose that $a' \equiv_C a$ with $a' \not\equiv_B a$. We must show that $|f(a')| < 1$. Choose a'' generically in U over $B^\ulcorner f^\urcorner aa'$. Then $a'' \equiv_B a$, so $|f(a'')| < 1$. Hence it suffices to show $|f(a')| \leq |f(a'')|$.

As $a' \not\equiv_B a$, there is a B-definable proper sub-ball V of U containing a'. Factorise f as $f(x) = d \prod(x - d_i)$. If $d_i \notin U$, then $|x - d_i|$ is constant for all $x \in U$, so $|a'' - d_i| = |a' - d_i|$. If $d_i \in V$, then $|a' - d_i| < |x - d_i|$ for all $x \in U \setminus V$, so $|a' - d_i| < |a'' - d_i|$. Suppose $d_i \in U \setminus V$. In this case, the generic choice of a'' forces $|a'' - d_i| \geq |a' - d_i|$. Considered together, these three cases ensure $|f(a')| \leq |f(a'')|$, as required.   □

EXAMPLE 15.11. Here J-independence holds but sequential independence fails. Let C be an algebraically closed valued field which is not maximally complete. Let $\{U_i : i \in I\}$ and $\{V_j : j \in J\}$ be chains of C-unary sets, chosen so that $U = \bigcap_{i \in I} U_i$ and $V = \bigcap_{j \in J} V_j$ are complete types. We may suppose that the cuts $\mathrm{rad}(U)$ and $\mathrm{rad}(V)$ in Γ are equal; also, that U and V are sufficiently independent that if $a \in U$ then no element of $\mathrm{acl}(Ua)$ lies in V, and likewise with U and V reversed. Let $b_1 \in U$, $b_2 \in V$. Choose $a_1 \in U$ generic over Cb_1b_2 and $a_2 \in V$ generic over $Cb_1b_2a_1$. Notice that $\Gamma(Ca_1a_2) = \Gamma(C) \neq \Gamma(Ca_1a_2b_1b_2)$. Then $a_1a_2 \perp_C^g b_1b_2$, but $|a_2 - b_2| > |a_1 - b_1|$ so $a_2a_1 \not\perp_C^g b_1b_2$. By Theorem 15.4, $a_1a_2 \perp_C^J b_1b_2$ and hence also $a_2a_1 \perp_C^J b_1b_2$.

EXAMPLE 15.12. This is an example where sequential independence holds but J-independence fails. Let C be any model. Let $p(x, y)$ be a type over C such that $p(x, y) \vdash |x| < 1 \wedge |y| < 1 \wedge |xy| = \gamma$, where $\gamma = |c|$ for some $c \in C$. Choose b generic in $\mathcal{M}$ and a_1, a_2 realising p such that $a_1a_2 \perp_C^g b$. The sequential independence implies that $|a_1| > |b|$, and since $|a_1a_2| = \gamma$, we have $|a_2| < \gamma|b|^{-1}$. Thus if $f(x) = c^{-1}bx_2$ then $f(x) \in J(a_1a_2/Cb) \setminus J((a_1a_2/C)$.

EXAMPLE 15.13. We give an example of a type with no J-independent extension over a certain set. Let C be a valued field, let γ lie in the Dedekind completion of $\Gamma(C)$ but not in $\Gamma(C)$, and let $a \in K$ with $|a| = \gamma$. Put

$A = C(a)$. Let $B = C(b_1, b_2)$, where $|b_1| > |b_2|$ and both $|b_1|$, $|b_2|$ lie in the same cut of $\Gamma(C)$ as γ. If $A \underset{C}{\overset{J}{\smile}} B$, then one cannot have $|a| < |b_1|$, for otherwise $b_1^{-1}x \in J(a/B) \setminus J(a/C)$. Thus $|a| \geq |b_1|$. Similarly $|a^{-1}| \geq |b_1|^{-1}$. Thus $|a| = |b_1|$. But similarly $|a| = |b_2|$, a contradiction. Thus, $\mathrm{tp}(A/C)$ has no J-independent extension over B.

REMARK 15.14. The proof in [12] of elimination of imaginaries of [12] can now be summarized as follows. First, for any stably dominated type q, the J-invariant $J(q)$ shows that the canonical base of q is coded. Indeed $J(q)$ is a submodule of a polynomial ring over K; hence it is canonically the union of R-submodules of finite-dimensional vector spaces, with canonical K-bases; one quickly reduces to sublattices, i.e., elements of S_n, and associated elements of T_n.

Finite sets and unary sets are dealt with separately. Given that, it remains to code functions f, on some neighborhood of each complete type over a set $C = \mathrm{acl}(C)$; in fact unary types suffice. If the type p is stably dominated, we first code the germ of f on p. Let q be the stably dominated type obtained by applying the function $x \mapsto (x, f(x))$ to p. Then the canonical base of q, coded above, serves also to code the germ of f. Since the germ is strong, there exists a C-definable function agreeing, on a neighborhood of the given type, with the original one.

If p is a limit of a Γ-family of stably dominated types, defined over C, we code the function restricted to each stably dominated type in the family using the previous step. Over a rich enough base, any type is a Γ-limit of stably dominated types; but such a family may not be defined over C. For 1-types however, there exists such a family whose *germ* is defined over C. Using our understanding of definable functions from Γ, we are able to deal with germs of maps from Γ and conclude the proof.

The appropriate higher-dimensional generalization of the fact quoted above about 1-types, over an arbitrary base, is not yet clear.

CHAPTER 16

OTHER HENSELIAN FIELDS

We give two examples of metastable theories other than ACVF. This means that much of the technology developed in the present manuscript applies to these theories. The main intention here is to illustrate the way that theorems proved in the ACVF context apply to other valued fields; all the real content of the lemmas below derives from ACVF theorems in the main text. At the end we make remarks concerning generalizations.

THEOREM 16.1. $\mathrm{Th}(\mathbb{C}((t)))$ *is metastable over the value group.*

Let $T = \mathrm{Th}(\mathbb{C}((t)))$. This theory admits elimination of valued field quantifiers, relative to Γ-quantifiers. It is simpler than other Hensel fields of residue characteristic 0 in that it also eliminates residue field quantifiers; this allows the statement below to have a simpler form, involving Γ alone, but a similar quantifier elimination involving k (or rather RV) is true in general. These observations can be found in [29].

Let $L \models T$. We view L as a subfield of a model M of ACVF. As L is Henselian, $\mathrm{Aut}(L^{\mathrm{alg}}/L) = \mathrm{Aut}_v(L^{\mathrm{alg}}/L)$, i.e., every field-theoretic automorphism of L^{alg}/L preserves the valuation. By Galois theory, $L = \mathrm{dcl}(L)$ within the field sort of M. We interpret the sorts S_n by $(S_n)_L = \mathrm{GL}_n(L)/\mathrm{GL}_n(R_L)$. We similarly interpret the sorts T_n though in the case of T they are redundant. In particular, $\Gamma_L = \{v(a): a \in L\}$. This is distinct from $\Gamma(L)$, but $\Gamma(L) = \Gamma_L \otimes \mathbb{Q}$ is the divisible hull of Γ_L.

Let $\mathcal{U}^T$ be a universal domain for T. We view it as a subset of a universal domain $\mathcal{U}$ for ACVF, interpreting the $\mathcal{G}$-sorts as above.

It can be shown with methods similar to those used here that T admits elimination of imaginaries in this language; but we will not require this fact.

Convention. By 'formula', 'definable function', 'type' we mean the quantifier-free ACVF notions; thus $\mathrm{tp}(a/C)$ denote the quantifier-free (ACVF) type of a over C. The corresponding T-notions are marked with T, e.g., $\mathrm{tp}_T(a/C)$ denotes the T-type; $\mathrm{Inv}_x^T(C)$ is the set of $\mathrm{Aut}(\mathcal{U}^T/C)$-invariant types of $\mathcal{U}^T$ in the variable x. We let $\Gamma^T = \Gamma_{\mathcal{U}^T}$, $S_n^T = (S_n)_{\mathcal{U}^T}$, and denote the field sort of $\mathcal{U}^T$ by $K(\mathcal{U}^T)$. For any $C \leq T$, let $\Gamma^T(C) = \Gamma(C) \cap \Gamma^T$. When $d \in \Gamma^T$, $\mathrm{tp}_T(d)$ denotes the set of Γ-formulas true of d in $\mathcal{U}^T$.

Let $\mathcal{F}$ be the collection of definable functions f on a sort of $\mathcal{G}$ into Γ^k, with the property that if $a \in \mathcal{U}^T$ then $f(a) \in \mathcal{U}^T$. Also let $\mathcal{F}(C)$ be the set of functions $f(x,c)$ with $f \in \mathcal{F}$ and $c \in C^m$. We can think of $\mathcal{F}$ as generating all definable functions into Γ, by virtue of:

LEMMA 16.2. *Let f be a definable function into Γ. Then for some $n, nf \in \mathcal{F}$.*

PROOF. Since the algebraic closure of $\mathcal{U}^T$ is a model of ACVF, and $\mathrm{acl}(\mathcal{U}^T) = \mathbb{Q} \otimes \Gamma^T$, it follows that any element of Γ definable over $\mathcal{U}^T$ lies in $\mathbb{Q} \otimes \Gamma^T$. Thus for any $a \in \mathcal{U}^T$, for some n, we have $nf(a) \in \Gamma^T$. The graph of nf is defined by a quantifier-free formula ψ_n, and we have shown that $\mathcal{U}^T \models (\forall x) \bigvee_{n \in \mathbb{N}} (\exists y) \psi_n(x,y)$. By compactness, for some finite set $n_1, \ldots, n_k \in \mathbb{N}$, $\mathcal{U}^T \models (\forall x) \bigvee_{i=1}^{k} (\exists y) \psi_{n_i}(x,y)$, i.e., some $n_i f(x) \in \mathcal{U}^T$ for any $x \in \mathcal{U}^T$. Let $n = \Pi_{i=1}^m n_i$. Then $nf(x) \in \Gamma^T$ for any $x \in \mathrm{dom}(f)$. So $nf \in \mathcal{F}$. $\square$

Let $C \leq \mathcal{U}^T$, $s \in S_n^T$ defined over C, and let $V = \mathrm{red}(s)$. Write V^T for the image of the $\mathcal{U}^T$-points under red.

Let x be a variable of one of the sorts $\mathcal{G}$. By a *basic formula* we mean one of the form $\psi(g(x))$, where $g \in \mathcal{F}$, and ψ is a formula of Γ^T. If $f, f' \in \mathcal{F}$ then $(f, f') \in \mathcal{F}$, so a Boolean combination of basic formulas is a basic formula.

LEMMA 16.3. (1) *Let $c, c' \in \mathcal{U}^T$, and assume $\mathrm{tp}_T(g(c)) = \mathrm{tp}_T(g(c'))$ for any $g \in \mathcal{F}$. Then $\mathrm{tp}_T(c) = \mathrm{tp}_T(c')$.*

(2) *Every formula in the sorts $\mathcal{G}$ is T-equivalent to a Boolean combination of basic formulas.*

(3) *For any $C \subseteq \mathcal{U}^T$ and $c, c' \in \mathcal{U}^T$, if $\mathrm{tp}_T(g(c)/C) = \mathrm{tp}_T(g(c')/C)$ for any $g \in \mathcal{F}(C)$, then $\mathrm{tp}_T(c/C) = \mathrm{tp}_T(c'/C)$.*

Equivalently, let $C' = C \cup \Gamma^T(C(c))$, and let $T_{C'}$ be the elementary diagram of C' in $\mathcal{U}^T$. Then $\mathrm{tp}(c/C') \cup T_{C'} \vdash \mathrm{tp}_T(c/C')$.

PROOF. (1) By relative quantifier elimination, this is true for formulas in the field sort. We will reduce to this case using resolution.

Conjugating by an element of $\mathrm{Aut}(\mathcal{U}^T)$, we may assume $g(c) = g(c')$ for any $g \in \mathcal{F}$. By Lemma 16.2, $\mathrm{tp}(c/\Gamma) = \mathrm{tp}(c'/\Gamma)$, and in particular $\mathrm{tp}(c) = \mathrm{tp}(c')$. We may assume $\Gamma(c) \neq (0)$; otherwise the hypothesis will apply to the tuples (c, t) and (c', t), while the conclusion for these tuples will be stronger. By Corollary 11.9 (applied to ACVF) there exists a sequence of field elements d such that $\mathrm{tp}(d/c)$ is isolated, $\Gamma(d) = \Gamma(c)$, and $c \in \mathrm{dcl}(d)$. Say $c = F(d)$, F an ACVF-definable function. Moreover $\mathrm{tp}(d/c)$ is realized in any $H = \mathrm{dcl}(H)$ containing representatives of the classes coded by c. It follows that d can be found in $\mathcal{U}^T$, and that there exists $d' \in \mathcal{U}^T$ with $\mathrm{tp}(d'c') = \mathrm{tp}(dc)$.

As in Corollary 11.16, we have $\Gamma(d) = \Gamma(c)$. Hence for any $g \in \mathcal{F}$ there exists a definable h into Γ with $g(d) = h(c)$ and $g(d') = h(c')$. Since $nh \in \mathcal{F}$ for some n, it follows that $g(d) = g(d')$. By the field case, we have $\mathrm{tp}_T(d) = \mathrm{tp}_T(d')$. Since $c = F(d), c' = F(d')$, we have $\mathrm{tp}_T(c) = \mathrm{tp}_T(c')$. This proves (1).

By (1), any T-type is determined by the basic formulas in it. (2) is a standard consequence, using compactness.

(3) It follows from (2), replacing variables by constants, that every formula over C is equivalent to a Boolean combination of basic formulas over C. So if $\text{tp}_T(g(c)/C) = \text{tp}_T(g(c')/C)$ for any $g \in \mathcal{F}(C)$, then $\text{tp}_T(c/C) = \text{tp}_T(c'/C)$. $\square$

If $q \in \text{Inv}_{xy}(C)$ and $p = q|x \in \text{Inv}_x(C)$, say q is orthogonal to Γ relative to p if for any $C' \supseteq C$, if $(c, d) \models q|C'$ then $\Gamma(C'c) = \Gamma(C'cd)$.

LEMMA 16.4. *Let* $P \in \text{Inv}_x^T(C)$, $r \in S_{xy}(C)$, $p \in \text{Inv}_x(C)$. *Let* $\tilde{p} = p|\mathcal{U}^T$, *and suppose* $\tilde{p} \subseteq P$. *Assume*: $p \cup r \vdash q$ *for some* $q \in \text{Inv}_{xy}(C)$, *and that* q *is orthogonal to* Γ *relative to* p. *Then* $P \cup r \vdash Q$ *for a unique* $Q \in \text{Inv}_{xy}^T(C)$.

PROOF. Let $C \subseteq C' \subset \mathcal{U}^T$, $c \models P|C'$, $(c, d) \models r$. We have to show that $\text{tp}_T(d/C'c)$ is determined. But by the relative orthogonality assumption, $\Gamma^T(C'cd) = \Gamma^T(C'c)$, so by Lemma 16.3, $\text{tp}(d/C'c)$ implies $\text{tp}_T(d/C'c)$. $\square$

COROLLARY 16.5. *Let* $C = \text{acl}(C) \cap \mathcal{U}^T$. *Then every* T-*type over* C *extends to an* $\text{Aut}(\mathcal{U}^T/C)$-*invariant* T-*type over* $\mathcal{U}^T$.

PROOF. First, it is an exercise (see also [20]) to check that any 1-type over C of T in the Γ-sort has an invariant extension.

We next argue that for a single field element a, $\text{tp}_T(a/C)$ has an $\text{Aut}(\mathcal{U}^T/C)$-invariant extension over $\mathcal{U}^T$. As for ACVF, if $(V_i : i \in I)$ is the set of C-definable closed balls which contain a, we say a is generic in $V := \bigcap(V_i : i \in I)$.

Suppose first that the chain has a least element, namely V, and that V is a closed ball; note that as the value group is discrete, there is no open/closed distinction for balls. Let $W = \text{red}(V)$ be the corresponding k-space. Let red: $V \to W$ be the natural map. Consider T-types $Q(x, y)$ with $\text{red}(y) = x$. Then the hypotheses of Lemma 16.4 apply, as $\{y : \text{red}(y) = x\}$ is also a closed ball so has generic orthogonal to Γ. Hence a complete C-invariant T-type is determined by (i) genericity of x in the affine k-space (ii) the formula $\text{red}(y) = x$ (iii) a complete type over C in x.

Suppose next that V is not a ball (so I has no least element) but has a C-definable point (or subtorsor) x_0 inside. Let $\gamma := |a - x_0|$. Extend first $\text{tp}(\gamma/C)$ to an invariant type, and then find the generic type of $B_{\leq \gamma}(x_0)$ using the closed ball case.

Finally, suppose that I has no least element and V has no C-definable ball. By the last case, for any $d \in V$ we described a $C(d)$-invariant type consisting of elements of V; it did not depend on the choice of d, so is C-invariant.

To complete the proof of the lemma, it suffices by Lemma 4.10 (i) to prove it for types of sequences of field elements. We allow infinite sequences, and reduce immediately to transcendence degree 1, i.e., to $\text{tp}(a, b/C)$ where a is a singleton, and b enumerates a part of $\text{acl}(C(a))$. Let $r = \text{tp}(a, b/C)$. Let P

be an $\mathrm{Aut}(\mathcal{U}^T/C)$-invariant extension of $\mathrm{tp}(a/C)$, and let p be the restriction to a quantifier-free ACVF-type over C. The conditions of Lemma 16.4 are met, so $P \cup r$ generates a complete T-type Q. Since r (being over C) and $P|\mathcal{U}^T$ are $\mathrm{Aut}(\mathcal{U}^T/C)$-invariant, so is Q.   □

LEMMA 16.6. *Let C be a maximally complete model of T, $C \prec M \models T$, and let a be a tuple from M. Let $C' = C \cup \Gamma^T(C(a))$. Then $\mathrm{tp}_T(A/C')$ is stably dominated.*

PROOF. Let $C^+ = C \cup \Gamma(L))$, and let L enumerate L. By Theorem 12.18 (i), $\mathrm{tp}(L/C^+)$ is stably dominated; by Remark 12.19, it is in fact stably dominated by a sequence $b \in \mathrm{dcl}(L) \cap \mathcal{U}^T$. Let p (respectively q) be the $\mathrm{Aut}(\mathcal{U}/C^+)$-invariant type extending $\mathrm{tp}(b/\mathrm{acl}(C^+))$ (respectively $\mathrm{tp}(L,b/\mathrm{acl}(C^+)))$; let $r = \mathrm{tp}(L,b/C^+)$. Then stable domination via b means that $r \cup p \vdash q$. Noting that b lies in the stable part of $\mathcal{U}^T$, let P be the $\mathrm{Aut}(\mathcal{U}^T/\mathrm{acl}^T(C^+))$-invariant type extending $\mathrm{tp}_T(b/\mathrm{acl}(C^+))$. By Lemma 16.4, $P \cup r$ generates an invariant type, extending $\mathrm{tp}(L,b/\mathrm{acl}(C^+)$, which is therefore stably dominated. Hence $\mathrm{tp}_T(L/C^+)$ is stably dominated. By Corollary 4.10 (ii), $\mathrm{tp}_T(a/C')$ is stably dominated.   □

PROOF OF THEOREM 16.1. Immediate from Corollaries 16.5 and 16.6, taking into account Corollary 4.12.   □

Generalizations. (1) Other value groups.
The only facts used about $\mathrm{Th}(\mathbb{Z})$ are:
  (a) Every type over a set C extends to an $\mathrm{Aut}(\mathcal{U}/C)$-invariant type.
  (b) If $A \prec B \models \mathrm{Th}(\mathbb{Z})$ then A/B is torsion free.
The class of value groups satisfying these include $\mathrm{Th}(\mathbb{Z}^n)$, and intermediate groups between $\mathbb{Z}$ and $\mathbb{Q}$. For applications the class seems comfortable.
(2) Valued fields of mixed or positive characteristic p, with p-divisible value groups satisfying (a,b). If the residue field remains algebraically closed, Theorem 16.1 goes through.
(3) Non-algebraically closed residue fields. This raises two issues.
  (a) The quantifier elimination is relative to higher congruence groups. In residue characterstic 0, we have quantifier elimination relative to $RV = K^*/(1+\mathcal{M})$, rather than to Γ. Nevertheless the method of proof generalizes. See [19] for related results, describing definable sets in terms of RV-definable families of ACVF-definable sets.
  (b) If the residue field is not stable, one cannot expect stably-dominated types; the notion needs to be refined.
  Let S be a partial type, $\mathcal{D}$ be a collection of sorts. Consider a $*$-definable map $f : S \to \mathcal{D}$; i.e., f is represented by a sequence $f_i : S \to D_i$ with $D_i \in \mathcal{D}$. Let I be an ideal on the (relatively) definable subsets of $f(S)$. Say S is dominated by $\mathcal{D}$ via (f, I) if for any definable set $X \subseteq S$, for all $d \in f(S)$ outside an I-small set, the fiber $f^{-1}(d)$ is disjoint from X or contained in X.

In the case of Henselian fields of residue characteristic 0, $\mathcal{D}$ will be the collection of definable sets internal to the residue field. When S is a complete stably dominated type, f will be the restriction of a definable function of ACVF. Therefore the Zariski closure of $f(S)$ has (in ACVF) finite Morley rank. The ideal I will consist of definable sets whose Zariski closure has lower dimension than $f(S)$.

In mixed characteristic $(0, p)$ one has quantifier elimination relative to the family of quotients $R/p^n R$. The p-adics present an interesting case. For a single such quotient, in the p-adic case, the ideal I is improper; but taken together the maps can be seen as having image in $\mathbb{Z}_p$, where the proper Zariski closed sets of n-dimensional space form a proper ideal. For a partial type S, one has a dominating map not into $\mathbb{Z}_p^n$ itself, but into a pro-definable set internal to $\mathbb{Z}_p^n$ in an appropriate sense (uniformly internal to the quotients $R/p^n R$). This of course connects to Pillay's idea of compact domination, providing instances of the phenomenon outside a group-theoretic setting.

Valued differential fields. Let $\widetilde{\text{VDF}}$ be the model completion of the theory VDF of valued differential fields ([47]). $\widetilde{\text{VDF}}$ extends ACVF and admits quantifier elimination. The residue field is stable as a sort in $\widetilde{\text{VDF}}$; it has the induced structure of a model of the theory DCF of differentially closed fields.

THEOREM 16.7. $\widetilde{\text{VDF}}$ *is metastable.*

PROOF. Let $\mathcal{U}$ be a universal domain for $\widetilde{\text{VDF}}$. Relying on Corollary 4.12, we will work without imaginaries. We first show that types extend to invariant types. Let L_v be the language of valued fields and $\mathcal{U}_v$ the restriction of $\mathcal{U}$ to L_v. Let $C \leq \mathcal{U}$, and let $p(x_0)$ be a type over C. Add variables x_i denoting $D^i x$, obtaining a type $P(x_0, x_1, \ldots)$. The new type P is generated by the formulas $Dx_i = x_{i+1}$, along with the restriction P_v of P to L_v. Let Q_v be an $\text{Aut}(\mathcal{U}_v/C)$-invariant L_v-type, extending P_v. Given a valued differential field C' extending C, let $C'' = C'(c_0, c_1, \ldots)$ be a valued field extension of C', generated by a realization of $Q_v|C'$. Then $C(c_0, c_1, \ldots)$ is linearly disjoint from C' over C. Therefore any derivation on $C(c_0, c_1, \ldots)$ extending the given derivation on C extends to a derivation on C''. It follows that $Q = Q_v \cup \{Dx_i = x_{i+1}\}$ is consistent. By quantifier elimination Q is a complete extendible type of $\widetilde{\text{VDF}}$. This shows that any type over C extends to an $\text{Aut}(\mathcal{U}/C)$-invariant type.

Similarly, let M be a model of $\widetilde{\text{VDF}}$, maximally complete as a valued field. For any $a_0 \in \mathcal{U}$, let $a_0, a_1, \ldots$ and P be as above. Let γ enumerate the value group of $M(a_0, a_1, \ldots)$, and let c enumerate the residue field of $M(a_0, a_1, \ldots)$. Then P is dominated by $\text{tp}(c/M, \gamma)$ over M, γ, in the sense of ACVF. It follows as in Lemma 16.6 that p is stably dominated. □

NOTE ADDED IN PROOF. Hrushovski has shown that the rigid analytic theories considered in [32] satisfiy condition (ii) of the definition of metastability (Definition 4.11). These are theories of expansions of a complete algebraically closed valued field by rings of separated power series. Condition (i) (existence of invariant extensions) hold at least when $C = \mathrm{acl}(C \cap K)$, i.e., the base is the algebraic closure (in all the sorts) of a *field*.

REFERENCES

[1] J. AX and S. KOCHEN, *Diophantine problems over local fields I*, **American Journal of Mathematics**, vol. 87 (1965), pp. 605–630.

[2] ———, *Diophantine problems over local fields II*, **American Journal of Mathematics**, vol. 87 (1965), pp. 631–648.

[3] ———, *Diophantine problems over local fields III*, **Annals of Mathematics**, vol. 83 (1966), pp. 437–456.

[4] J. BALDWIN, **Fundamentals of stability theory**, Perspectives in Mathematical Logic, Springer, Berlin, 1988.

[5] J. BALDWIN and A. LACHLAN, *On strongly minimal sets*, **The Journal of Symbolic Logic**, vol. 36 (1971), pp. 79–96.

[6] W. BAUR, *Die Theorie der Paare reell abgeschlossner Körper*, **Logic and algorithmic (in honour of E. Specker)**, Monographie No. 30 de L'Enseignement Mathématique, Université de Genève, Geneva, 1982, pp. 25–34.

[7] S. BUECHLER, **Essential stability theory**, Springer–Verlag, New York, 1996.

[8] Z. CHATZIDAKIS and E. HRUSHOVKI, *Model theory of difference fields*, **Transactions of the American Mathematical Society**, vol. 351 (1999), pp. 2997–3051.

[9] O. ENDLER, **Valuation theory**, Springer, Berlin, 1972.

[10] Y. ERSHOV, *On elementary theories of local fields*, **Algebra i Logika**, vol. 4 (1965), pp. 5–30.

[11] B. HART, *Stability theory and its variants*, **Model theory, algebra, and geometry** (D. Haskell, A. Pillay, and C. Steinhorn, editors), Mathematical Sciences Research Institute Publications, vol. 39, Cambridge University Press, Cambridge, 2000, pp. 131–149.

[12] D. HASKELL, E. HRUSHOVSKI, and H.D. MACPHERSON, *Definable sets in algebraically closed valued fields: elimination of imaginaries*, **Journal für die Reine und Angewandte Mathematik**, vol. 597 (2006), pp. 175–236.

[13] D. HASKELL and H. D. MACPHERSON, *Cell decompositions of C-minimal structures*, **Annals of Pure and Applied Logic**, vol. 66 (1994), pp. 113–162.

[14] J. E. HOLLY, *Canonical forms of definable subsets of algebraically closed and real closed valued fields*, **The Journal of Symbolic Logic**, vol. 60 (1995),

pp. 843–860.

[15] E. HRUSHOVSKI, *Valued fields, metastable groups*, preprint.

[16] ——, *Unidimensional theories are superstable*, **Annals of Pure and Applied Logic**, vol. 50 (1990), no. 2, pp. 117–138.

[17] ——, *The Mordell–Lang Conjecture for function fields*, **Journal of the American Mathematical Society**, vol. 9 (1996), pp. 667–690.

[18] ——, *Stability and its uses*, **Current developments in mathematics, 1996 (Cambridge, MA)**, International Press, Boston, 1997, pp. 61–103.

[19] E. HRUSHOVSKI and D. KAZHDAN, *Integration in valued fields*, **Algebraic geometry and number theory: In honor of Vladimir Drinfeld's 50th birthday** (V. Ginzburg, editor), Progress in Mathematics, vol. 253, Birkhauser, 2006, pp. 261–405.

[20] E. HRUSHOVSKI and B. MARTIN, *Zeta functions from definable equivalence relations*, preprint.

[21] E. HRUSHOVSKI, Y. PETERZIL, and A. PILLAY, *Groups, measures and NIP*, **Journal of the American Mathematical Society**, (to appear), math.LO/0607442.

[22] E. HRUSHOVSKI and A. TATARSKY, *Stable embeddedness in valued fields*, **The Journal of Symbolic Logic**, vol. 71 (2006), pp. 831–862.

[23] A. A. IVANOV and H. D. MACPHERSON, *Strongly determined types*, **Annals of Pure and Applied Logic**, vol. 99 (1999), pp. 197–230.

[24] A.A. IVANOV, *Strongly determined types and G-compactness*, **Fundamenta Mathematicae**, vol. 191 (2006), pp. 227–247.

[25] T. JECH, **Set theory**, Academic Press, New York, London, 1978.

[26] I. KAPLANSKY, *Maximal fields with valuations I*, **Duke Mathematical Journal**, vol. 9 (1942), pp. 303–321.

[27] ——, *Maximal fields with valuations II*, **Duke Mathematical Journal**, vol. 12 (1945), pp. 243–248.

[28] B. KIM and A. PILLAY, *Simple theories*, **Annals of Pure and Applied Logic**, vol. 88 (1997), pp. 149–164.

[29] F-V. KUHLMANN, *Book on valuation theory*, in preparation, http://math.usask.ca.

[30] S. LANG, **Algebra**, Addison–Wesley, Reading, Sydney, 1971.

[31] D. LASCAR, *The category of models of a complete theory*, **The Journal of Symbolic Logic**, vol. 47 (1982), pp. 249–266.

[32] L. LIPSHITZ, *Rigid subanalytic sets*, **American Journal of Mathematics**, vol. 115 (1993), pp. 77–108.

[33] A. J. MACINTYRE, *On definable subsets of the p-adic field*, **The Journal of Symbolic Logic**, vol. 41 (1976), pp. 605–610.

[34] A. J. MACINTYRE, K. MCKENNA, and L. VAN DEN DRIES, *Elimination of quantifiers in algebraic structures*, **Advances in Mathematics**, vol. 47 (1983), pp. 74–87.

[35] H. D. MACPHERSON and C. STEINHORN, *On variants of o-minimality*, *Annals of Pure and Applied Logic*, vol. 79 (1996), pp. 165–209.

[36] D. MARKER, *Introduction to the model theory of fields*, **Model theory of fields** (D. Marker, M. Messmer, and A. Pillay, editors), Lecture Notes in Logic, vol. 5, Springer, Berlin, 1996, pp. 1–37.

[37] ———, **Model theory. An introduction**, Graduate Texts in Mathematics, vol. 217, Springer, New York, 2002.

[38] P. M. NEUMANN, *The structure of finitary permutation groups*, **Archiv der Mathematik**, vol. 27 (1976), pp. 3–17.

[39] A. PILLAY, *Forking, normalisation, and canonical bases*, **Annals of Pure and Applied Logic**, vol. 32 (1986), pp. 61–81.

[40] ———, **Geometric stability theory**, Oxford Logic Guides, vol. 32, Clarendon Press, Oxford, 1996.

[41] A. PILLAY and G. SROUR, *Closed sets and chain conditions in stable theories*, **The Journal of Symbolic Logic**, vol. 49 (1984), pp. 1350–1362.

[42] B. POIZAT, **Groupes stables**, Nur al-Mantiq wal-Ma'riah, 1987, Translated as **Stable groups**, American Mathematical Society, Providence, RI, 2001.

[43] ———, **A course in model theory. An introduction to contemporary mathematical logic**, Springer, New York, 2000.

[44] P. RIBENBOIM, **The theory of classical valuations**, Springer, Berlin, 1998.

[45] A. ROBINSON, **Complete theories**, North–Holland, Amsterdam, 1956.

[46] T. SCANLON, personal communication.

[47] ———, *A model complete theory of valued D-fields*, **The Journal of Symbolic Logic**, vol. 65 (2000), no. 4, pp. 1758–1784.

[48] S. SHELAH, **Classification over a predicate. II. Around classification theory of models**, Lecture Notes in Mathematics, vol. 1182, Springer, Berlin, 1986.

[49] ———, **Classification theory and the number of nonisomorphic models**, second ed., Studies in Logic and the Foundations of Mathematics, vol. 92, North–Holland Publishing Co., Amsterdam, 1990, xxxiv+705 pp.

[50] ———, *Quite complete real closed fields*, **Israel Journal of Mathematics**, vol. 142 (2004), pp. 261–272.

[51] ———, *Dependent first order theories, continued*, **Israel Journal of Mathematics**, (to appear).

[52] G. SROUR, *The notion of independence in categories of algebraic structures, Part I: basic properties*, **Annals of Pure and Applied Logic**, vol. 38 (1988), pp. 185–213.

INDEX

1-torsor
  special, 84

acl-generated, 91
algebraic closure, 19
atomic, 28

ball, 77
BS, 66
  for ACVF, 106

canonical base, 26
  dcl, 40
code
  strong, 63
conjugate, 31

dcl-generated, 91
definable, 19
definable closure, 19
divides, 24
domination-equivalence, 22

equivalence relation
  finite, 25
  opaque, 119
  opaquely layered, 121

forks, 24
function
  $*$-function, 41

κ-generated, 66
generic, 86
  basis, 89
  in value group, 86
  resolution, 89
  type, 86
generically independent, 87
geometric sorts, 78
germ, 63
good separated basis, 127
group scheme, 123, 164

Henselian, 75
henselisation, 75

i-complete, 147
i-domination, 150
i-extension, 147
imaginaries, 21
  code for, 21
  elimination of, 20
immediate extension, 75
independence
  domination $\underset{}{\downarrow}{}^{d}$, 34
  generic, $\underset{}{\downarrow}{}^{g}$, 87
  J-independence, $\underset{}{\downarrow}{}^{J}$, 165
  modulus, $\underset{}{\downarrow}{}^{m}$, 156
  non-forking, $\underset{}{\downarrow}$, 24
  sequential, $\underset{}{\downarrow}{}^{g}$, 92
  stationary, $\underset{}{\downarrow}{}^{s}$, 37
indiscernible, 28, 36
  in ACVF, 115
internal, 30
  to the residue field, 82
invariant extension, 35
invariant extensions
  in ACVF, 93–96
invariant type, 21
  and generic independence, 87
  and stable domination, 52
  and stationary independence, 38

J-independence, 165

$\Lambda(s)$, 78
lattice, 78

maximally complete, 75, 127
metastable, 55
minimal, 28
modulus independent, 156
Morley sequence, 29

Neumann's Lemma, 57

o-minimal, 18
1-module, 83
1-torsor, 83
opaque
  equivalence relation, 119
  group, 121, 124
opaquely layered, 120
order-like, 88
orthogonal
  and stable domination, 111
  to residue field, 131
  to value group, 87, 107

place, 74
positive type, tp$^+$, 155
pre-resolution, 119
preweight, 26
prime, 28
pseudo-convergent sequence, 75
Puiseux series, 76

radius of subtorsor, 83
rank
  Cantor–Bendixson, 20
  Δ-rank, 23
red(T), 84
residue field, 74
resolution, 119
  canonical open, 109
  dcl, 122
  generic closed, 108
  prime, 119, 124
resolved, 108

saturated, 20
sequential independence, 92
S_n, 78
St, 31
stable
  formula, 22
  set, 30
  theory, 22
stable domination, 34
  in ACVF, 97–103
  over algebraic closure of base, 42
  transitivity, 69
  via a $*$-function, 41
stable, stably embedded, 30
Stably dominated
  for invariant types, 36
stably embedded, 29
  residue field, 77

value group, 77
stationary independence, 37–41
strong code, 63
  on a definable type, 63
  on a stably dominated type, 64
strongly minimal, 27
substructure, 81
subtorsor, 83
Swiss cheese, 85
  block, 85
  hole, 85
  trivially nested, 85

T_n, 78
torsor, 83
  subtorsor, 83
totally transcendental, 27
type
  $*$-type, 41
  base for, 22
  definable, 23
  Δ-type, 23
  extendible, 23
  invariant, 21
  isolated, 19
  Lascar strong, 25
    in ACVF, 99
  stationary, 25, 104
  strong, 25
  unary, 83

unary
  code, 85
  sequence, 84
  set, 83
  type, 83
unary set
  special, 84
unary transfinite sequence, 92

valuation
  non-trivial, 73
valuation ring, 73
value group, 73
  $\Gamma(C)$, 81
  Γ_C, 81
vi-complete, 147
$VS_{k,C}$, 82

weight, 26